Rivers and Lakes

In the same series

Birds
Wild Flowers
Mushrooms and Fungi
Aquarium Fishes
Minerals and Rocks
Insects
Fishes
Poisonous Plants and Animals
Water and Shore Birds
Trees and Shrubs

Chatto Nature Guides

Rivers and Lakes

Plants and animals illustrated and identified with colour photographs

Eckart Pott

Translated and edited by
Gwynne Vevers

Chatto & Windus · London

Published by
Chatto & Windus Ltd.
40 William IV Street
London WC2N 4DF
*
Clarke, Irwin & Co Ltd.
Toronto

ISBN 0 7011 2543 8 (Hardback)
ISBN 0 7011 2544 6 (Paperback)

Title of the original German edition:
Bach-Fluss-See

Printed in Germany

Introduction

The waters of the earth are all involved in a gigantic circulatory system, powered by the sun as an energy source. A small part (0.0008%) is in the form of water vapour in the atmosphere. This water reaches the earth as rain or snow. There some of it evaporates and re-enters the atmosphere. Most of the earth's water (83.51%) is in the oceans, about 15% is held in the ground, while just over 1% is held as ice in the glaciers and polar regions. Finally, the rivers and lakes have only 0.015% of the earth's waters, so these inland waters have only a minute amount compared with the seas.
The seas represent the oldest living environment on earth, whereas most lakes are relatively young, having been formed after the last ice age—a short time span within the framework of evolutionary history. The seas represent the "cradle of life" and they have an amazing wealth of plant and animal species. Furthermore, the marine environment provides very constant living conditions over long periods of time. Inland waters were colonized from the seas, and they have a much smaller range of living orgnisms. They are also subject to much greater seasonal and daily fluctuations than the seas.

Types of inland waters
There are several types of inland water. Some are surface waters, others subterranean. The latter include underground caves which in spite of the unusual conditions contain a number of characteristics organisms, such as the Cave Slater (*Asellus cavaticus*) and the Olm (*Proteus anguinus*), an amphibian found in the subterranean waters in the mountains of Yugoslavia. Ground water held in the soil also has specially adapted organisms.
Surface waters can be divided into flowing and standing waters. Flowing waters include springs, mountain and lowland streams and rivers.

Standing waters include natural and artificial lakes, ponds and pools. Lakes are characterized by having a shallow shore zone or littoral, usually colonized by plants, and a deeper zone, the profundal, in which plant growth is impossible owing to lack of light. Dams can be regarded as artificial lakes. Ponds can be defined as lakes with shallow water, scarcely exceeding 3m in depth. Here one cannot speak of a littoral and a profundal, as the whole of the bottom may be colonized by plants. Artificial ponds may form when a stream is dammed, the water level often being controlled by a weir. Small bodies of water which often dry up in summer may be called pools, and then there are numerous even smaller collections of water, such as puddles, water-filled hollow trees and certain plant organs.
Finally, inland salt waters should also be mentioned, as they are quite characteristic. They are found primarily in arid zones, where the rate of evaporation exceeds the rate of precipitation.

Flowing waters

In these waters the important factor is the current, to which the organisms must be adapted. If they are not they may be killed or may drift away. Two types of current can be distinguished. In laminar currents the individual components move one above the other in the general direction of flow. In turbulent currents the components compete with each other. Both types of current may occur in a stream or river. Laminar currents occur particularly in places near the bottom, over rocks or other objects lying in the water, and also in clumps of algae and moss. In other parts turbulent currents will predominate.
Plants and animals living in flowing waters have to become adapted to the currents, and this they do in different ways. Some algae grow in the form of flat layers which lie close to the substrate, others produce thin filaments which stream out in the current. The algae of flowing waters are, of course, always fixed very firmly to the substrate
Animals avoid areas with great turbulence. Thus, they usually live on the underside of rocks, particularly by day. Many move out at night on to the upper surfaces where they hunt for food. Certain animals possess morphological adaptations that help them to withstand water currents. In some the shape is considerably modified, as in the River Limpet (*Ancylastrum*

fluviatile). Many aquatic insect larvae have a very flat body which allows them to live on rocks in places with laminar currents, e.g. the larva of the mayfly *Epeorus*. Many insect larvae have developed various ways of attaching themselves. Thus, mayfly larvae of the genus *Rhithrogena* have a kind of suction disc on the ventral surface. The larva of the small midge *Liponeura* has six large ventral suction discs. Other larvae attach themselves firmly to the substrate by claws on the feet or by hooks on the abdomen.

However, in spite of these mechanisms animals in flowing waters do tend to drift, particularly in the evening when they move out on to the upperside of rocks in search of food. During the course of the night the rate of drift decreases.

In addition to the currents, the temperature naturally has a considerable effect on organisms in flowing waters. The temperature is most constant both daily and seasonally in the area where streams arise—the source area. Further downstream the daily and annual fluctuations gradually increase. In general, the same applies to the content of carbon dioxide and dissolved substances.

Freshwater fishes also vary in their spawning habits, temperature preferences and oxygen requirements. In cold mountain streams with a high oxygen content the Salmonidae are quite characteristic, in particular the Brown Trout (*Salmo trutta fario*) and further down the Grayling (*Thymallus thymallus*). Even further downstream the river may have Barbel (*Barbus barbus*) and Common Bream (*Abramis brama*). If the river flows into the sea the brackish-water zone will have Flounders (*Platichthys flesus*). From source to estuary a river can therefore have regions characterized by Brown Trout, Grayling, Barbel, Bream and Flounder. Alternatively, a river can be subdivided according to the diagram on the following page.

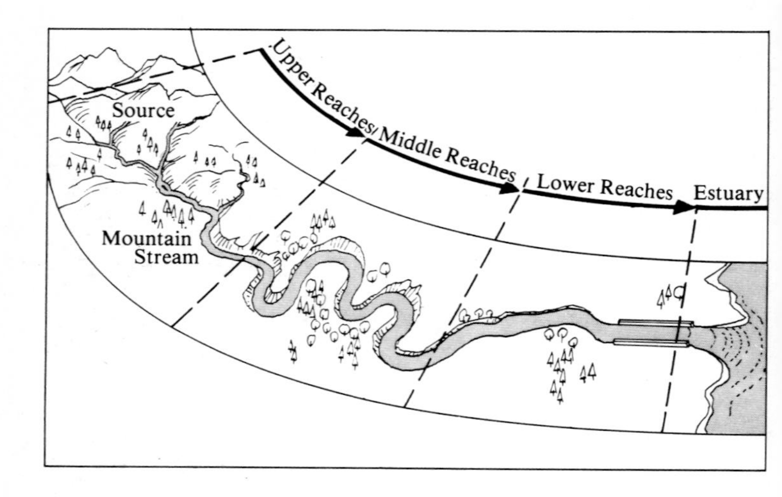

Standing Waters
Lakes are perhaps the best investigated of all inland waters. They have three principal habitats. The shore region or littoral is the transitional area between land and water and it is subject to changes in water level and to considerable fluctuations in temperature. On the other hand, the open water or pelagial shows less fluctuation. This is the habitat of the plankton (see below). The deep water or profundal is mainly without light and it offers the most constant environmental conditions.
A lake is, however, subject to considerable seasonal fluctuations. In summer one can speak of stagnation with water layers at different temperatures. Like the air above it the upper water layer, known as the epilimnion, is warm. Below it is a transitional layer or thermocline, the mesolimnion, in which the temperature falls rapidly. Finally the deep water is cold, usually about 4°C, a temperature at which water has its greatest density. In autumn the surface waters cool off until all the water in the lake is at a uniform temperature and the stratification disappears, a process that is helped by autumnal storms.
In winter the surface water becomes much colder and the deep water is now warmer than the surface. This is the stage of

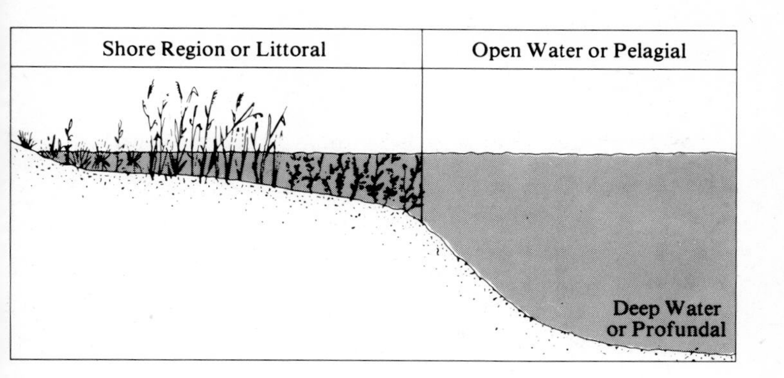

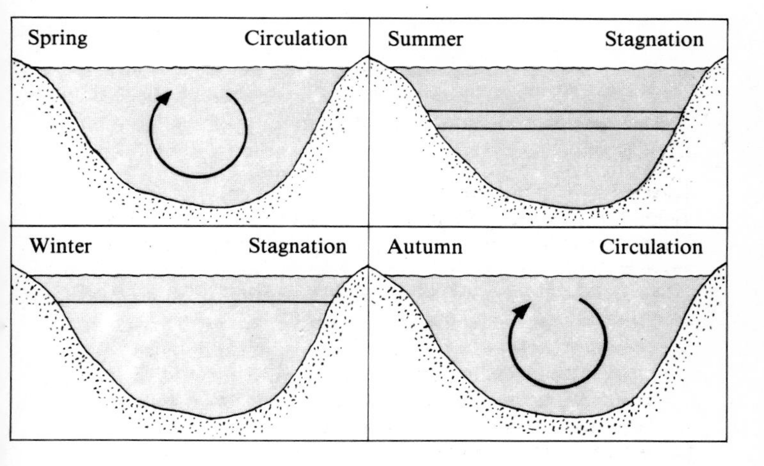

winter stagnation. In spring the upper waters again warm up slowly, the winter stratification disappears and the total waters circulate. Finally, the stable summer stratification is re-formed. However, not all lakes have these annual cycles of circulation and stagnation. In some lakes the water only becomes stratified once in the year. In certain cases only parts of a lake may become stratified.

Lakes can also be characterized on other criteria, such as the intensity of their biological productivity. A lake with poor production is known as oligotrophic, one with good

production as eutrophic. There are transitional stages which can be termed mesotrophic. Moorland areas with poor drainage may have distrophic lakes in which the conditions are largely unsuitable for larger plants and animals. This is why certain of the Scottish lochs are so relatively barren.
The term eutrophication is used when an originally oligotrophic or mesotrophic lake becomes eutrophic owing to the inflow of extra nutrients, resulting in an increase in biological production. Such enrichment may proceed faster than the rate of decomposition of organic matter and when this happens a layer of foul detritus accumulates on the bottom, which is poor in oxygen but often contains sulphur compounds, giving conditions that are highly unsuitable for living organisms.

Marginal Vegetation

The edges of eutrophic lakes usually show characteristic zones of vegetation. Moving from the land towards the open water there may first be a marshy zone usually with Alder trees (*Alnus*) which passes to an area of shrubs, mainly willow (*Salix* spp.) and Alder Buckthorn (*Rhamnus frangula*). There is then a broad zone characterized particularly by various sedges (*Carex* spp.). Here the ground may appear to be dry for months at a time, but there is always water around the roots. Further towards the lake is the very typical reed-bed, with the Common Reed (*Pragmites communis*) in particular, but usually only a few other plant species. In areas rich in nutrients, as in the vicinity of an estuary, reedmaces *Typha* spp. may completely replace the Common Reed. Open areas in the reed-bed may have bur-reeds (*Sparganium* spp.), Mare's-tail (*Hippuris vulgaris)*, Arrowhead (*Sagittaria sagittifolia*) and the Common Water Plantain (*Alisma plantago-aquatica*). In relatively peaceful waters Bulrushes (*Scirpus lacustris*) may appear as outposts of the reed-bed. This plant can also grow underwater and can therefore penetrate into deeper water than the Common Reed. As the depth increases there will be rooted plants with leaves that float at the surface, such as the White Waterlily (*Nymphaea alba*) and Yellow Waterlily (*Nuphar lutea*) and pondweeds (*Potamogeton* spp.). Various other aquatic plants are completely submerged, with only their flowers extending above the water surface. Finally, in deeper water there are certain algae and in particular the species of *Chara*.

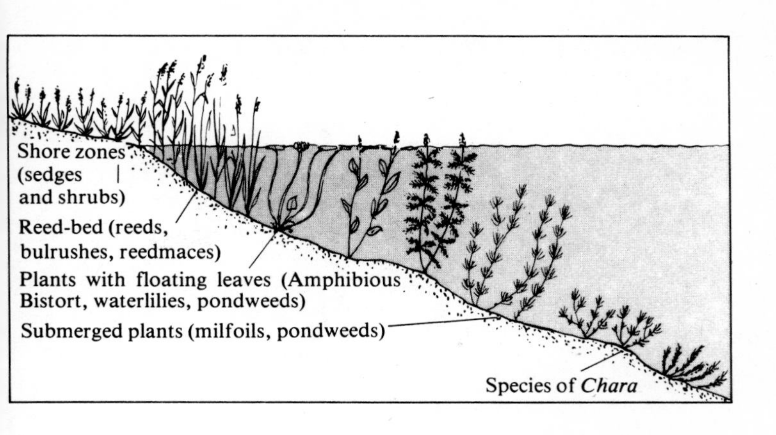

The plankton

The pelagial region of a lake may appear as a monotonous habitat, but in fact it contains a wealth of floating organisms, known collectively as the plankton. The plant plankton or phytoplankton consists of various microscopic algae which grow and reproduce in the presence of sunlight, certain dissolved nutrients (nitrates, phosphates, etc.) and the carbon dioxide dissolved in the water. The process involved is known as photosynthesis. It is important that phytoplankton plants should remain as long as possible close to the surface, for only there is the light sufficient for photosynthesis. Only a few of these tiny plants can move actively and the majority rely on various methods to reduce their rate of sinking. Although they are heavier than water they can gain buoyance by having gas bubbles. In fact many algae contain gas vacuoles within their cells which are often so effective that when productivity is high, there may be so-called water blooms with the algae extending across the surface in large masses. Other algae surround their cells with lightweight jelly, or they may contain fat globules which increase buoyancy. Certain algae are shaped so that they remain more buoyant. Thus some are in the form of a disc or a tiny starlet, so that they tend to float in the water rather like a parachute in the air.

The animal plankton or zooplankton in fresh waters is much poorer in species than that of the oceans. It consists mainly of rotifers or wheel animalcules (Rotifera) and various small

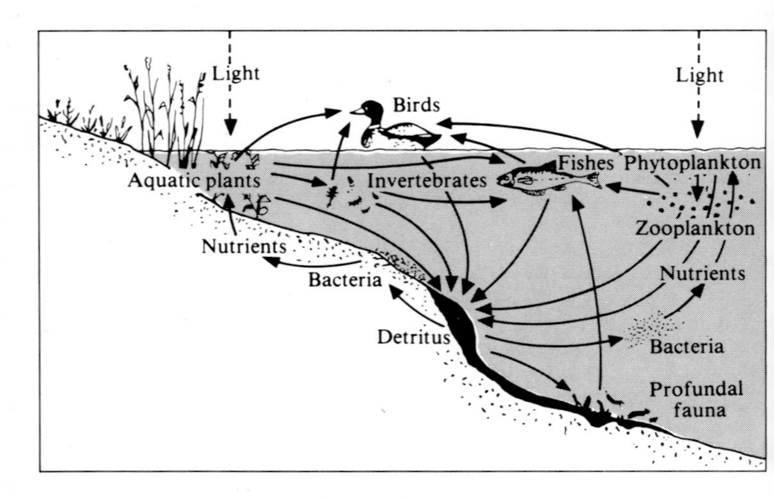

crustaceans. Only a few other animal groups have planktonic forms, e.g. the phantom larvae of *Chaoborus* and the larva of the Zebra Mussel (*Dreissena polymorpha*). These and the small crustaceans can be seen with the naked eye, but the rotifers require a microscope.

Rotifers have an organ which creates a current that wafts microscopic algae towards them. Most water-fleas and other small creatures feed in a similar way. Some forms are predatory, feeding on other zooplankton animals. The zooplankton is itself eaten by various fish species, such as young Perch, Three-spined Sticklebacks and Whitefish. The diagram above shows some of the links in a typical freshwater food chain.

Rivers and Lakes

Common Reed *Phragmites communis*

A cosmopolitan species, widely distributed in Europe in swampy areas and particularly along the edges of standing and slow-flowing waters. This is one of the most characteristic plants in the reed-bed zone of lakes and dams, extending out into water to a depth of 1.50 metres, where it gives place to underwater plants, including those with floating leaves.

The Common Reed is a perennial grass with a much branched rootstock from which the tall shoots (1-4m) arise. As in all grasses the stems are extremely flexible. The pointed, lanceolate leaves have a long sheath. The flowers are carried in panicles which are up to 40 cm long and slightly drooping with numerous individual flowers. In Europe, flowering occurs in August-September. The seeds do not ripen until the following January, but this does not always happen, and the plant is usually propagated asexually from the rootstock or by long surface runners.

This plant not only determines the appearance and structure of reed-beds as a habitat, but also changes these, contributing considerably to the consolidation of the banks. Large amounts of plant detritus and mud accumulate around the stems, sometimes in sufficient amounts to allow land plants to grow. If this process continues a body of standing water may gradually disappear, leaving an area of flat marshy ground. Reeds are also important ecologically as they provide spawning sites for amphibians and fishes, and also nesting sites for various birds and particularly warblers. Below the water surface the stems become overgrown with algae on which snails and other aquatic animals feed.

During recent years it has been noticed that in some places areas of reed-bed have died out. Several reasons for this have been suggested including water pollution and grazing by swans, but the subject is still being investigated.

Great Reedmace *Typha latifolia*

(above, left)

An abundant plant in Britian, growing to a height of 1-2 m, along the edges of lakes, ponds and ditches. The leaves are long and linear and the flowers are carried in a dense, cylindrical spike. Often erroneously called Bulrush. The Lesser Reedmace (*T. angustifolia*) is shorter and not so widely distributed in Britain.

Bulrush *Scirpus lacustris*

(above, right)

An almost cosmopolitan sedge, widely distributed in Britain along the banks of pond, lakes and ditches, often extending beyond the reed-bed zone. The short, creeping rootstock produces stiff, erect stems up to 3 m tall. The flowers, which are arranged in spikelets, appear in June-July.

Branched Bur-reed *Sparganium erectum*

(below, left)

A member of the reedmace family distributed over most of the northern hemisphere, including Britain, but not in the tropics. The rootstock is creeping and the stems, up to 80 cm tall, carry long, linear leaves. The flowers are in the form of spherical heads, the upper ones being male, the lower female. The plant grows along the margins of lakes, ponds and slow-flowing streams, and it flowers in June-August.

Common Sedge *Cladium mariscus*

(below, right)

A widely distributed sedge in Europe, extending northwards to southern Scandinavia, but local in England and rare in Scotland. The creeping rootstock produces rush-like, leafy stems 80-200 cm in height with flowers arranged in numerous clusters. The leaves are very rough and the edges are armed with tiny sharp hooks which probably protect the plants from being eaten. They grow in ditches, marshland and fens, particularly in areas with hard water.

Common Water Plantain
Alisma plantago-aquatica (above, left)

A perennial plant distributed throughout Europe, temperate Asia and North America, and abundant in Britain, particularly in the shallow, muddy parts of lakes and ponds and also in slow-flowing streams. The long-stalked, ovate to lanceolate leaves arise from a thick rootstock, and grow in the form of a rosette. The small, pale pink flowers are in loose, pyramidal panicles and they appear in June-September. The flower stems may reach a height of 1 m.

Arrowhead *Sagittaria sagittifolia*
(above, right)

Distributed throughout temperate Europe and Asia, including Britain, in areas of shallow flowing and standing water with a muddy bottom. This is a perennial plant with a creeping rootstock which forms bulb-like tubers in the autumn. In spring the linear submerged leaves appear first, then long-stalked floating leaves and finally the characteristic arrow-shaped aerial leaves. The erect flower-stem appears in June-August and carries whorls of large white flowers, the upper ones being male, the lower female.

Water Horsetail *Equisetum fluviatile*
(below, left)

Distributed throughout Europe, including Britain, northern Asia and North America, where it grows along the edges of lakes at altitudes up to 2400 m. The long creeping rootstock gives off shoots up to 1.20 m in height. The terminal fruiting bodies appear in May-June.

Sweet Flag *Acorus calamus* (below, right)

Originally native to eastern Asia, and believed to have been introduced into Europe as a medicinal plant. It grows in the shallow water areas of lakes and ponds, often in large numbers. The plant has a creeping rootstock and long, linear and erect leaves. The flowering stems appear in June-July, and bear flowers which are densely packed in cylindrical spikes up to 8 cm long.

Yellow Iris *Iris pseudacorus*

Distributed throughout Europe, parts of temperate Asia and North America, and abundant in Britain. It grows in marshy places, ditches and along the banks of lakes and streams, sometimes in large masses.

The linear, stiff and erect leaves are 1-3 cm broad and may reach a length of 50-100 cm. They grow from a thick horizontal rootstock which is often much branched. The flowering stems appear in May and June and are usually slightly taller than the leaves. The large flowers are a bright satiny yellow, with the typical iris shape. These produce green capsules containing numerous pale brown seeds.

The related Stinking Iris or Gladdon (*Iris foetidissima*) is a rather smaller plant with narrower leaves and a deeper green. The whole plant produces a disagreeable smell when bruised. The violet-blue or more rarely pale yellowish-white flowers appear in early summer. They produce capsules with bright orange or scarlet seeds. This is quite a common plant in western Europe, particularly in shady places. It is abundant in southern England, but scarcer in the north.

Water Soldier *Stratiotes aloides*

Widely distributed throughout Europe and temperate northern Asia, except in the extreme north; common in parts of England. It grows mainly in standing or slow-flowing waters, particularly in the shallows of lakes and ponds. The broad-linear leaves may reach a length of up to 41 cm and a width of 4 cm; they have sharply toothed edges and grow in funnel-shaped rosettes, below which is a dense mass of aquatic roots. The flowers are white, the male ones being separate from the female. These appear from May to July. This plant is also propagated asexually by runners formed in the leaf axils. These elongate and form leaves and roots at the end.

It is particularly interesting that during the winter this plant grows in the mud on the bottom, but in early spring floats to the surface. After the flowering period the plant again sinks down into the mud, where it overwinters.

Mare's-tail *Hippuris vulgaris*

Distributed throughout Europe, central Asia, North America and in other areas, such as Chile. It occurs in parts of Britain, growing in shallow lakes and ponds and in watery ditches, particularly in relatively cool, hard water. Depending upon the locality it may grow as a fully submerged plant or half-submerged, and the shoots may reach a length of up to 200 cm. They carry whorls of 6-12 narrow, linear leaves, those above water protruding horizontally from the stem. The whole plant is anchored in the bottom by a perennial, creeping rootstock. The flowers are minute and inconspicuous, appearing in the axils of the aerial leaves, and consisting of a single stamen and a single ovary. Mare's-tail should not be confused with the horsetails (see p. 18) which are not flowering plants.

Marsh Marigold *Caltha palustris*

Distributed throughout Europe, temperate Asia and North America, this is a very common plant in Britain, growing along the edges of streams and marshy places up to altitudes of 2000 m. It has leaves below and above the water surface. The flowers appear in March or April, sometimes later. The flowers consist of five yellow sepals (there are no true petals), numerous stamens and 5-10 carpels. Pollination is by bees and flies, sometimes also by small beetles, which are attracted by the nectar glands in the flower. After pollination each carpel produces several seeds. The whole plant grows from a perennial rootstock firmly anchored in the bottom. The erect, annual shoots are hollow and 15-40 cm in height. The leaves, which are hairless and kidney-shaped or heart-shaped, are carried on long stalks. The edges of the leaves have rounded notches.

Bog Arum *Calla palustris*

Widely distributed throughout Europe and parts of North America and temperate Asia, and introduced into Britain. It grows along the edges of ponds and lakes, in woodland swamps, in alder thickets and often in deserted peat workings. The distribution is somewhat local but in areas where it does occur the Bog Arum covers extensive areas. The plant overwinters in the form of a stout rootstock, from which the erect long-stalked leaves grow up in spring. The leaves are rounded or heart-shaped and they are unspotted. The flowers consist of a short, greenish spadix enclosed within a white sheath known as a spathe, and therefore similar in shape to those of the related Cuckoo-pint or Lords and Ladies (*Arum maculatum*). In the latter the leaves are often spotted. The seeds are in the form of red berries, which are poisonous.

White Waterlily *Nymphaea alba*

A widely distributed plant in Europe, including Britain, northern and central Asia, northern Africa and north-western America, growing mainly in fairly shallow standing waters, sometimes in slow-flowing rivers. It is a characteristic plant of the zone with plants that have floating leaves, in depths of not more than 1.5 m. It has a thick, creeping rootstock which may grow to a length of up to 1 m. It produces two different leaf types. The underwater leaves are thin, whereas the conspicuous floating leaves are oval and leathery with long stalks. The large white flowers lie on the surface of the water and are 5-20 cm in diameter. They have four green sepals, numerous spirally arranged white petals and a large number of bright yellow stamens. These conspicuous flowers appear from June to September. The closely related *Nymphaea candida* is a smaller plant with deeper yellow stamens and petals that are shorter than the sepals.

Fringed Waterlily *Nymphoides peltata*

A plant found in many parts of central and southern Europe, extending eastwards through temperate Asia to Japan, and growing in some parts of southern England. It grows usually in standing or very sluggish waters. The rootstock is long and creeping and the floating leaves are remarkably similar to those of the White Waterlily although the two plants are quite unrelated. The flowers are yellow, about 3 cm in diameter, and in the form of a five-lobed funnel; the edges are slightly fringed. In most areas flowering takes place between July and September. The seeds are ovate, but the plant is also propagated asexually from the rootstock.

Yellow Waterlily *Nuphar lutea*

(above, left)

A close relative of the White Waterlily (both belonging to the family Nymphaeaceae), with a similar wide distribution, and particularly common in the muddy parts of ponds and lakes. It sometimes extends further out into the lake than the White Waterlily, even to depths of 2 m. The rootstock, which may be up to 10 cm thick and 3 m long, produces relatively few thin underwater leaves and large, oval floating leaves up to 40 cm across. The yellow flowers appear in June-September.

Amphibious Bistort
Polygonum amphibium (above, right)

Widely distributed as a perennial in ponds and slow-flowing rivers at altitudes up to 2000 m throughout Europe, including Britain, temperate Asia and northern America. It overwinters as a thick creeping rootstock. The oblong or lanceolate leaves float at the surface and are leathery and hairless, but the plant can also grow on land and the leaves are then longer, narrower and slightly hairy. Both forms produce spikes of small pink flowers in June-September.

Broad-leaved Pondweed
Potamogeton natans (below)

The genus *Potamogeton* has a large number of species distributed throughout the world, and there are probably at least 20 species in Britain, and possibly several hybrids. *P. natans* has floating leaves, but some species are entirely submerged. Most live in the zone beyond the reeds, with other submerged plants, sometimes with waterlilies. The upper, floating leaves are ovate, thick and opaque and usually 6-8 cm long, whereas the submerged are thinner and narrower. The flowering spikes grow up above the water surface, mostly in May to September, the individual flowers being small and green, without true petals but with four sepals.

Lesser Duckweed *Lemna minor*

(above, left)

A cosmopolitan floating plant without true leaves or stem, but consisting of a single oval thallus, 2-3 mm in diameter, with only one tiny root. The related Greater Duckweed (*Spirodela polyrhiza*), which is 5-8 mm across, has a cluster of roots under each thallus. The tiny Rootless Duckweed (*Wolffia arrhiza*) is only 0.5-1.5 mm across. Duckweeds form a thin green carpet at the surface. They do possess tiny flowers without petals, but also reproduce vegetatively by budding off tiny new thalli.

Frogbit *Hydrocharis morsus-ranae*

(above, right)

A widely distributed plant in Europe, including Britain, and temperate Asia, growing in ponds and ditches or in the quieter parts of lakes. It prefers soft water, and so is often found in abandoned peat workings and other moorland waters. In contrast to the waterlilies this is a completely floating plant without a rootstock, and it derives its nutrients directly through tufts of unbranched aquatic roots. The stalked, roundish leaves, up to 4 cm in diameter, are heart-shaped at the base. The plant also has floating stems which produce further rosette-like plants, forming a dense carpet at the water surface. The flowers have three pale green outer segments and three white inner segments. They appear in May-August.

River Water Crowfoot
Ranunculus fluitans

(below)

Distributed throughout Europe, including Britain, northern Africa and temperate North America, mainly in rivers and streams. The plant produces submerged stems which may be up to 6 m long. The leaves are finely cut with thin, parallel segments. The flowers appear above the water surface between June and August. They are up to 2 cm in diameter, with five green sepals and five petals. They may not always produce seeds, owing to the uncertainty of pollination by insects. Certain other water crowfoots have floating leaves.

Water Fern *Azolla filiculoides*

This fern came originally from the warm temperate and subtropical parts of America. It has appeared in Europe as a greenhouse and aquarium plant, whence it has escaped and established itself in certain areas, including southern England. It has, for instance, become quite a serious weed in the rice-growing areas of southern France, and would not thrive except in a reasonably warm environment. Like Water Velvet (below) it floats at the surface. The much-branched stem bears pairs of very small leaves which overlap each other like the tiles on a roof. They are unwettable. There are also clusters of roots which hang down into the water and take up nutrients. If the waters dry up the fern produces more roots with which it takes nutrients from the ground. Parts of the plant contain cells of the blue-green alga *Anabaena azollae.* These fix atmospheric nitrogen which thus becomes available to the fern.

Water Velvet *Salvinia natans*

A small fern distributed from western Europe through Asia to Japan. It prefers standing and slow-flowing waters, particularly those that are relatively hard. The whole plant floats free at the surface. The branching stems may grow to a length of 20 cm. They carry pairs of small oval leaves c. 1 cm long, which are hairy on the upperside and unwettable. The leaves also contain air spaces which helps them to remain buoyant. On the underside of the floating leaves there are tufts of colourless submerse leaves which function as roots, and produce the spores. This is a warmth-loving plant, sometimes kept in the aquarium where it requires plenty of light.

Crystalwort *Riccia fluitans*

This is a liverwort with a cosmopolitan distribution; liverworts are non-flowering plants related to mosses. The plant consists of numerous branched thalli (not strictly leaves) which float just below the water surface in lakes, ponds and ditches, often forming dense mats. It can also grow on damp, muddy ground where it forms rosettes that are anchored by thin root-like structures. The floating aquatic form does not have these structures. The dense mats of Crystalwort tend to prevent light penetrating and they thus have an adverse effect on the growth of other aquatic plants, including some of the algae which form an early link in the food chain. On the other hand, Crystalwort releases much oxygen into the water.

Water Moss *Fontinalis antipyretica*

A true moss distributed throughout Europe from the lowlands up to the tree limit in the Alps, and common in Britain. In contrast to Crystalwort this is particularly a plant of small streams and other running waters, although in suitable places it also grows on the bottom of lakes. It is in fact a highly adaptable plant and its form varies considerably according to the environmental conditions. The plant grows in large clusters, with stems up to c. 50 cm in length. The broadly ovate leaves are pointed and sharply keeled and they are arranged in three rows along the stem. When growing in running waters this moss provides relatively good shelter against the current for many small invertebrates. It is always interesting to examine a sample of this moss under a lens to see the variety of life it contains.

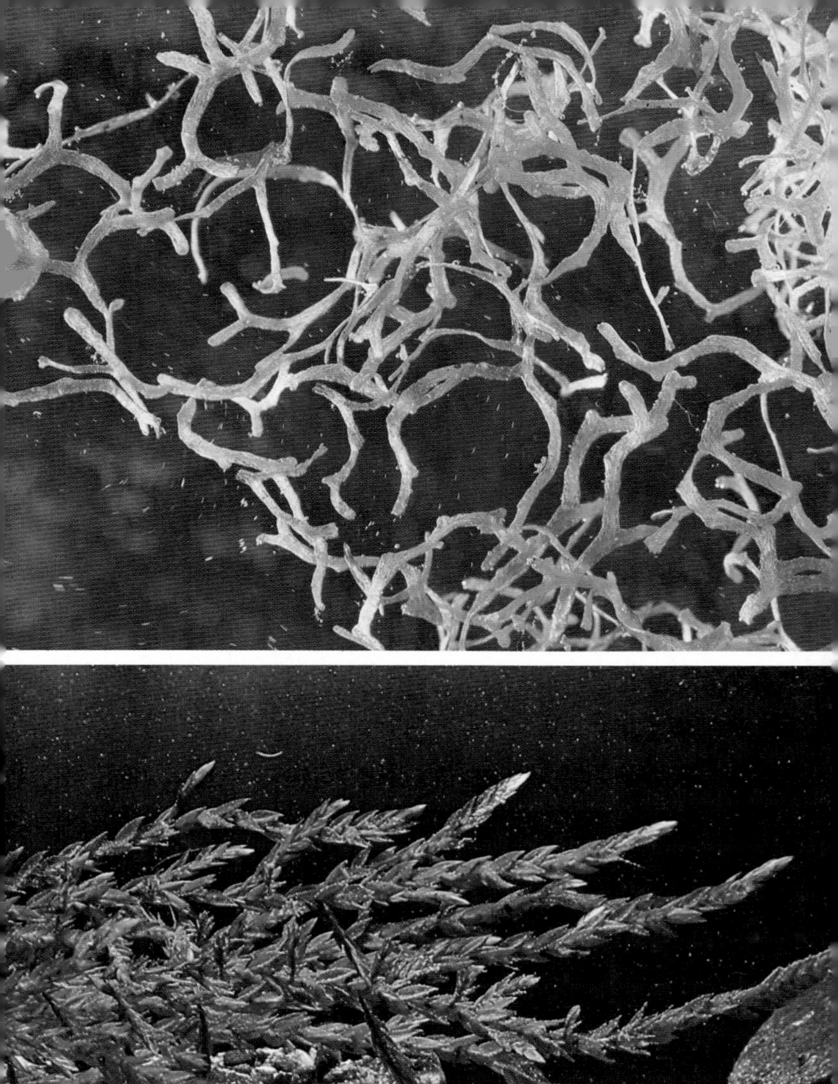

Water Violet *Hottonia palustris*

A perennial member of the Primrose family, found in many parts of Europe and not uncommon in southern Britain. It grows in ponds, lakes and canals, but does not extend into mountainous areas. The creeping rootstock produces numerous divided leaves which are all submerged, and it may grow to a length of almost 1 m. At intervals the plant sends up aerial shoots which carry the flowers in whorls of three to six. The flowers are white or pale pink with five petals, and they appear between May and July. After pollination the flowering stems bend over into the water and the numerous seeds ripen underwater.

Bladderwort *Utricularia* species

There are several species in Europe, including Britain, and also in other parts of the world. The European bladderworts are all floating plants with thin submerged leaves interspersed with small bladders or vesicles which are air-filled (below, right). These bladders have small bristles which act as triggers. When a small aquatic invertebrate, such as a water-flea, touches the trigger, the valve of the bladder opens and the water-flea is sucked in. The animal dies inside the bladder and is digested, the resulting material being taken up by the plant. Between June and September the plant sends up leafless flowering stems which each carry a few large yellow flowers (below, left). In Britain the most widely distributed species is the Greater Bladderwort (*Utricularia vulgaris*). The very similar Lesser Bladderwort (*U. minor*) has smaller, paler flowers. Both species occur in Europe, temperate Asia and North America.

Whorled Milfoil
Myriophyllum verticillatum (above, left)

Widely distributed throughout the northern hemisphere, in standing or slow-flowing waters, particularly those that are rich in nutrient, but not too hard. The plant is perennial, the rootstock being anchored in the mud at the bottom. The much-divided leaves are all submerged and arranged in whorls of four or sometimes five. The erect flowering stems appear from June to September, and the flowers are small and inconspicuous. The similar Spiked Milfoil (*M. spicatum*) is also widely distributed in Europe.

Holly-leaved Naiad *Najas marina*
(above, right)

The genus *Najas* has many species throughout the world and the present one is found in Europe mainly in lakes, more rarely in flowing waters; it occurs very rarely in Britain. It grows in depths down to c. 3 m, preferably in hard water, also in brackish water, and in some lakes forms green meadows on the bottom. The thin, much-branched stems produce small roots at the nodes or joints and these help the plant to spread across the bottom. The thin, linear and toothed leaves are usually arranged in whorls of three. The small, inconspicuous flowers appear in the leaf axils between June and September.

Stoneworts *Chara* species (below)

These are non-flowering plants which mostly occur in fresh waters. They belong to a highly developed family group of algae or near-algae and could at first sight be confused with flowering plants. The stems are up to 20 cm long, without numerous branches, covering large areas on the bottom. In general, the species of *Chara* usually extend to greater depths than the aquatic flowering plants.

Blue-Green Algae Cyanophyta

Like the bacteria, the blue-green algae do not possess a true cell nucleus (i.e. one enclosed in a membrane), which is a characteristic feature of all other plants. In the cells of blue-green algae it is usually possible to distinguish, under the microscope, an inner, colourless area which contains the hereditary material and an outer coloured area which contains the substances involved in photosynthesis, particularly chlorophyll a. The typical blue-green colour is due to certain bluish pigments. Chlorophyll a does not occur in photosynthetically active bacteria, which assimilate with the help of other pigments. Blue-green algae do not have chloroplasts, the coloured bodies seen in the cells of higher plants. In general, this group stands intermediate between the bacteria and the true algae.

Blue-green algae may occur as single cells or as filaments. The upper photograph shows a water bloom caused by the mass proliferation of *Microcystis aeruginosa.* This blue-green alga occurs in the form of microscopic spherical cells which often lie, thousands together, in poorly defined gelatinous coverings. The latter lower the specific gravity of the cells, a mechanism which prevents the mass from sinking in the water. In addition, the cells have gas vacuoles so that the mass rises to the surface where it spreads out. When conditions are favourable, *Microcystis* may proliferate so much that every millilitre of water has a distinct blue-green colour. The whole surface of a lake can then look like a thick green soup. Nevertheless, a water bloom does not necessarily indicate that the conditions in a lake are out of balance. *Microcystis* can occur in lakes that are only slightly polluted.

The lower photograph shows a filamentous blue-green alga *Oscillatoria limosa* as seen under the microscope. Such forms and also species in the genera *Nostoc, Anabaena* and *Spirulina* may float freely on account of their gas vacuoles or they may form dense mats on rocks or similar substrates.

Blue-green algae are widely distributed and may even occur in extreme habitats, such as in hot thermal springs or on glacier ice. They play the part of pioneers in the colonization of unoccupied habitats.

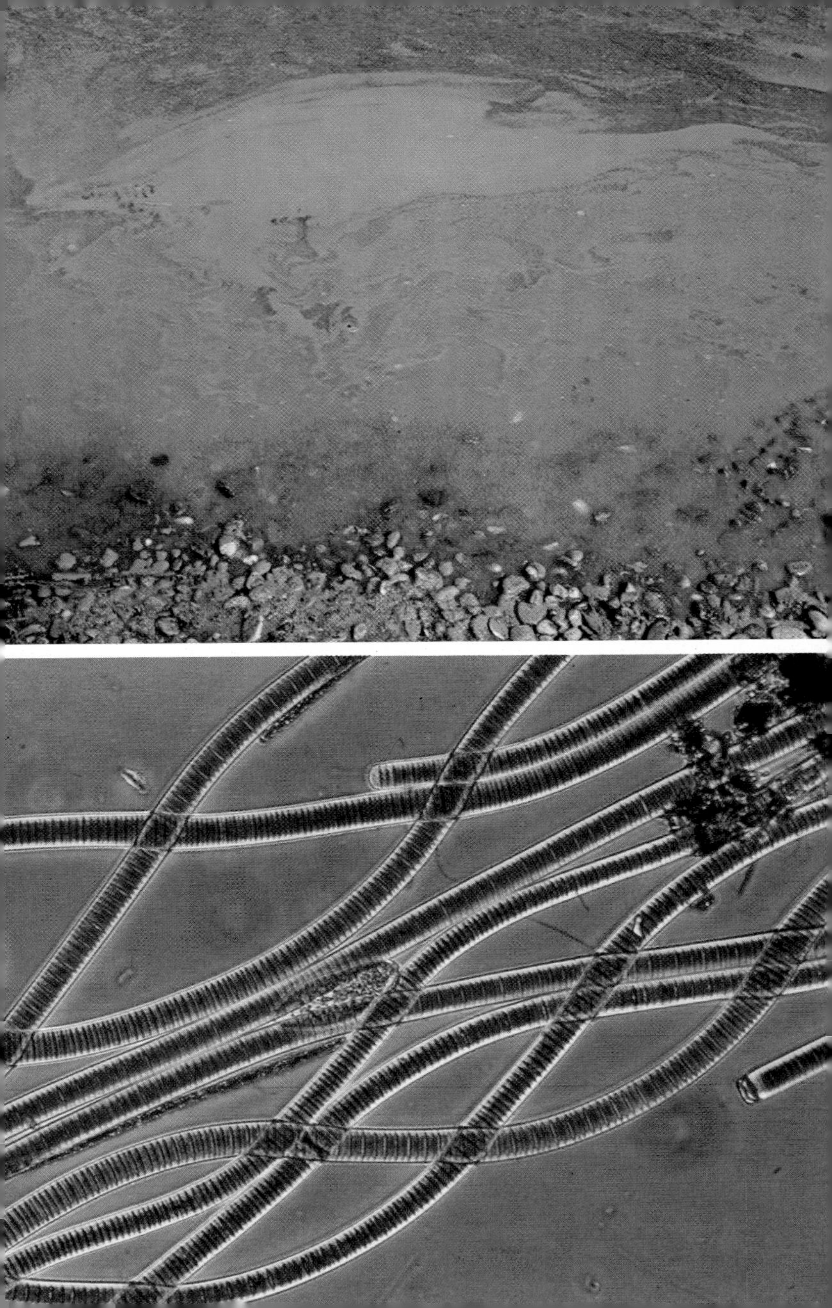

Blanket-weed *Cladophora* species

Felt-like green mats can often be found on rocks in rivers. Sometimes these may be blue-green algae (p. 40), but more usually they are filamentous green algae. Various forms can be distinguished. In cool streams rich in oxygen one frequently finds species of the genus *Ulothrix* which can easily be recognized under the microscope by its ring-like chloroplasts. Another common genus is *Cladophora* (below, left) the branched filaments of which often form dense masses in rivers and lakes. The chloroplasts lie in a layer close below the cell surface. Some species of *Cladophora* are so prolific that they may form conspicuous mounds or bunches of filaments in the water. The bunches may be torn off by wind and wave action and be driven ashore, as seen in the upper photograph.

As in all submerged plants, algal growth is dependent upon the content of nutrients in the water. In many places filamentous algae have increased in amount when the water has been enriched by pollution, so that it changes from being oligotrophic (poor in nutrients) to eutrophic (rich in nutrients). Such mass proliferation of an alga may be quite natural but nowadays it is often the result of human activities.

Spirogyra *Spirogyra* species (below, right)

This is a very common alga with long unbranched filaments, which forms extensive mats like those of *Cladophora*. Species of *Spirogyra* have a complicated sexual reproduction in which two filaments lie close together and there is an exchange of material between neighbouring cells. They also multiply vegetatively by the simple division of cells. There are more than 100 species of *Spirogyra* in Europe, mostly in standing water, but some occur in streams and rivers.

Scenedesmus species (above, left)

Several species of this planktonic alga occur in fresh waters, and almost every net haul of phytoplanton in a lake will bring up some specimens, which can then be examined under a microscope. They all have a stout cell membrane and are usually seen as 2-8 single cells joined together.

Volvox *Volvox globator* (above, right)

There are several species of *Volvox* in European waters. The one shown (under the microscope) occurs commonly in lakes rich in nutrient. It is in the form of a colony of about 1000 individual cells, connected to one another by bridges of protoplasm. Each cell has two flagella and these help to move the whole sphere. A small number of the cells take part in asexual reproduction, forming daughter spheres which grow within the parent sphere, which is c. 2 mm across. When the latter dies the daughter spheres are released. *Volvox* also has a form of sexual reproduction.

Diatoms Diatomeae (below, left)

Diatoms are common algal forms in fresh waters and in the sea. Some are sessile. living fixed to rocks or water plants, others float in the plankton. All diatoms are characterized by having each cell enclosed in a silica case. Species of the genera *Fragillaria, Asterionella* and *Diatoma* are common in lake phytoplankton.

Desmids Desmidiaceae (below, right)

These are amongst the most beautiful of all algae, but only of course when seen under the microscope as in the photograph of *Micrasterias* (below, right). Desmids are only to be found in fresh waters, particularly in moorland waters poor in nutrients. Other common forms belong to the genera *Closterium, Staurastrum* and *Cosmarium.*

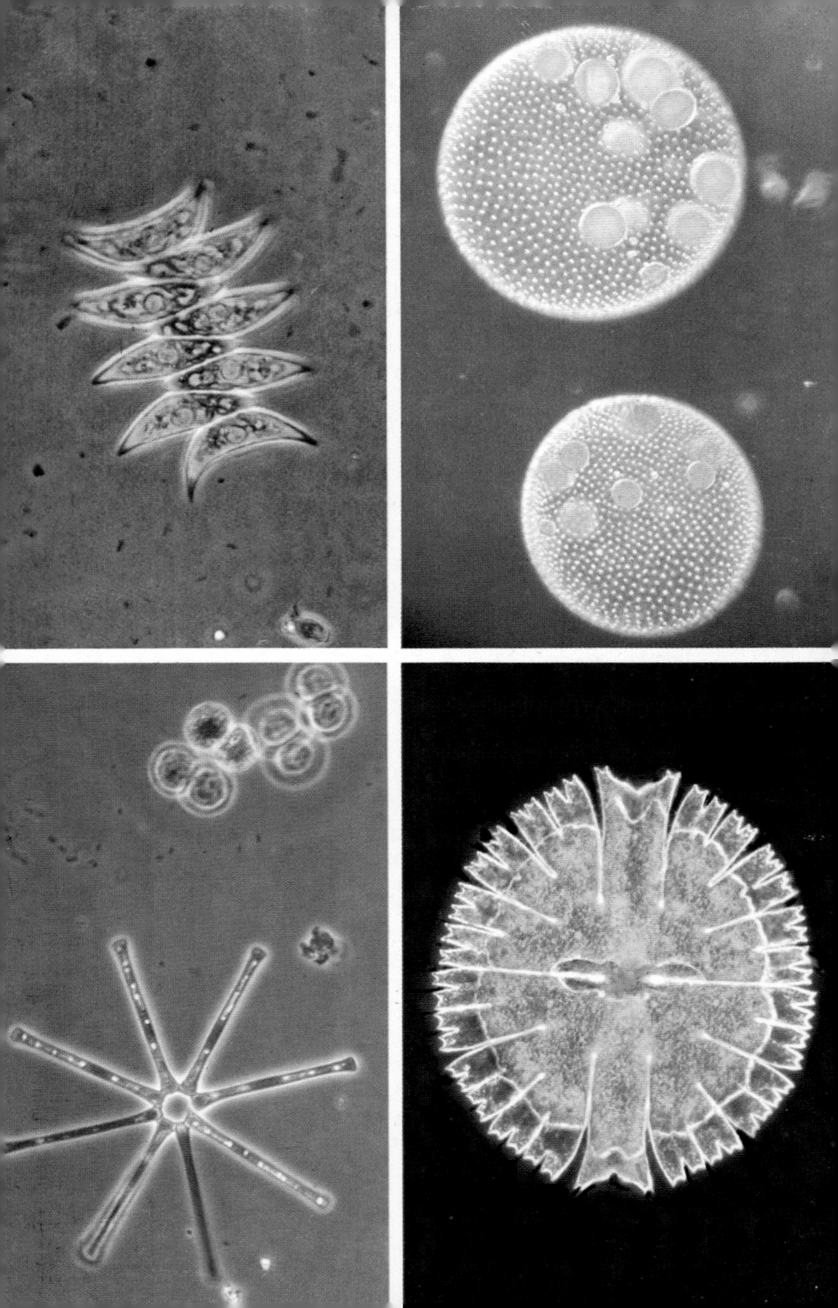

Brown Hydra *Hydra oligactis* (above, left)

Like the sponges the hydrozoans occurring in fresh waters have evolved from marine forms; the latter include hydroids and medusae. In fresh waters there are only about a dozen species classified in five genera. Hydras live in standing and flowing waters, particularly those rich in vegetation. They attach themselves to plants by the basal disc and extend the long tentacles armed with sting cells. With these they catch prey, such as water fleas, water mites, sometimes even fish larvae. The Brown Hydra is up to 3 cm long.

Freshwater Sponge *Spongilla lacustris* (above, right)

European fresh waters only have 5-6 species of sponge, a tiny number compared with the wealth of species living in the sea. The present species occurs as a yellowish-white or brownish encrustation on rocks, water plants, mollusc shells and submerged tree roots, in both standing and flowing waters. In general, the exact identification of these colonial animals is only possible with the aid of a microscope. These sponges overwinter as spherical granulae, from which a new sponge develops in the spring.

Rotifers Rotatoria (below, left and right)

Rotifers are often known colloquially as wheel animalcules. They are classified in three groups, of which the first is restricted to the sea. The second group live on firm substrates in fresh water and the third form part of the freshwater plankton. Rotifers, which are 0.1 to 2.5 mm long, can be caught with a fine mesh net. Some of the common species belong to the genus *Brachionus* (below, left). The characteristic wheel organ can be seen when the animal is examined under the microscope. This organ serves to move the rotifer and also to waft towards it a current of water containing plankton, particularly planktonic algae, on which the animal feeds. *Brachionus* has a firm outer covering, in contrast to the predatory *Asplanchna* (below, right) whose sac-like shape changes with every movement. This rotifer is also very transparent. It preys mainly on smaller rotifers.

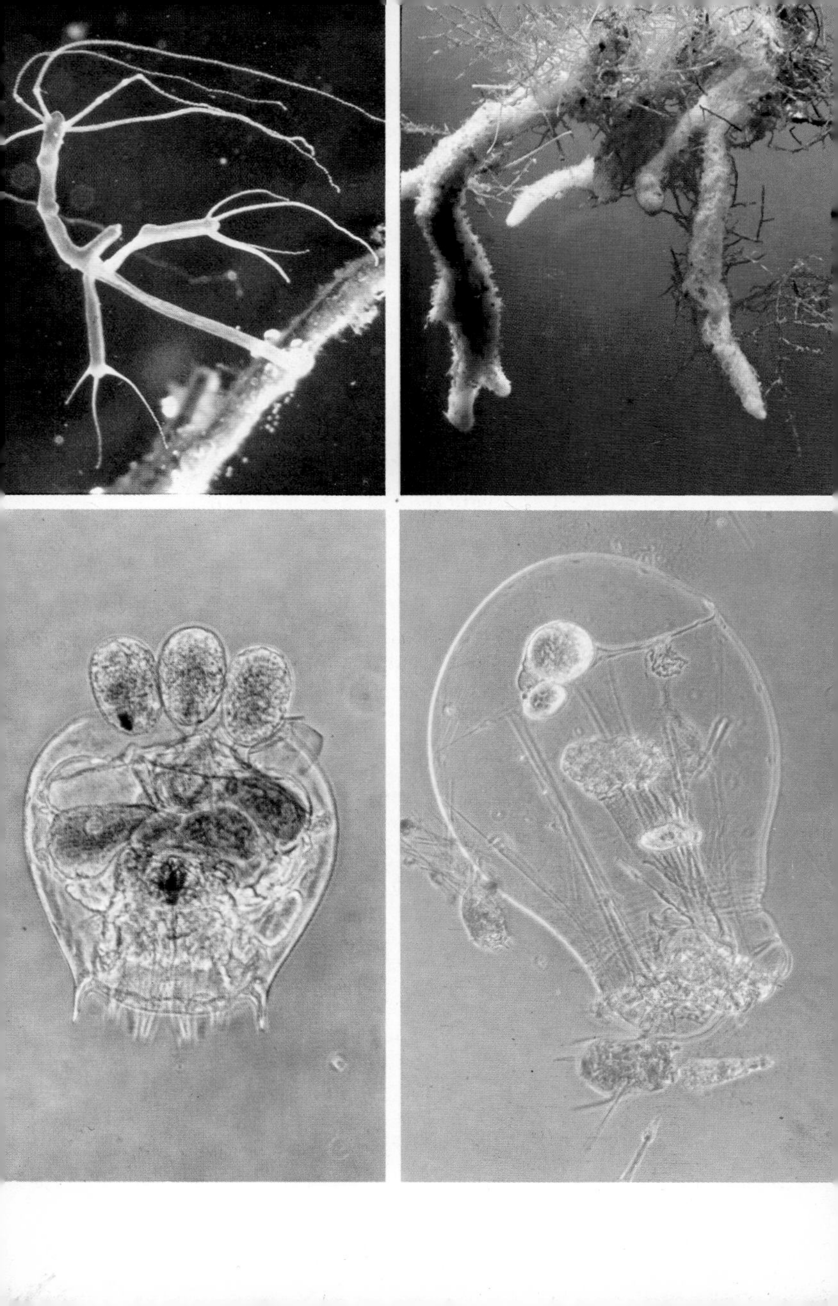

Pearl Mussel *Margaritifera margaritifera*

A widespread bivalve throughout most of the northern hemisphere, including Britain, but its numbers have, in some areas, become much reduced owing to increased water pollution. It is particularly associated with clear streams with soft water running over sandstone or igneous rocks, e.g. in southern Scotland. In the interests of its conservation, stocks are being transplanted from threatened waters to suitable, unpolluted rivers. This bivalve can be distinguished by the longish, kidney-shaped outline of the shell, usually with a very slight indentation of the lower edge. It is also characteristic that the hinge region is much abraded. The shell reaches a length of up to 12 cm, and is dark brown to black depending upon the age of the animal. The interior of the shell is covered with an iridescent layer of mother-of-pearl. True pearls are in fact found in some specimens and these were at one time the subject of commercial fisheries.

Swan Mussel *Anodonta cygnea*

Widely distributed in many parts of Europe, including Britain, in both standing and flowing waters. This is probably the commonest freshwater mussel in the area, and there are several local varieties. The shell has a longish-ovate outline and is much wider than that of the Pearl Mussel. It is also larger and lengths of up to 20 cm have been recorded. However, the most important identification character is the absence of teeth at the hinge. This differentiates the Swan Mussel, not only from the Pearl Mussel, but also from the similar freshwater mussels of the genus *Unio*.

Zebra Mussel *Dreissena polymorpha*

This species originally came from the area of the Caspian and Black Seas, whence it has spread across the inland waters of Europe. It is now not uncommon in rivers, canals and lakes in Britain, where it was first recorded in 1824. Its populations have evidently fluctuated considerably in some European waters.

Externally the shell somewhat resembles that of the Common Mussel (*Mytilus edulis*) found so commonly on the seashore. The shell is shiny yellow-brown marked with dark brown stripes, and is usually 2.5 to 4 cm in length. The animal attaches itself to rocks and other submerged objects by a bunch of sticky threads known as the byssus. In some places water-pipes have become blocked by the settlement of groups of Zebra Mussels. The eggs and sperm are shed at random into the water where fertilization takes place. The fertilized eggs produce larvae which live for some weeks floating in the plankton where they develop into miniature mussels which already have a pair of tiny shells. It is at this stage that the mollusc becomes attached. It usually reaches the adult size in 1-3 years.

Orb-shell Cockle *Sphaerium rivicola*

These are small bivalve molluscs, common in many inland waters in Europe, including Britain. The shell is up to 27 mm in length. This species is widespread in canals and slow-flowing rivers. The other common species is *Sphaerium corneum* which is only c. 11 mm long. It can be found on the bottom among gravel, but also among the stems of water plants. A third species *S. lacustre* is typically found in stagnant waters.

The pea-shell cockles of the genus *Pisidium* are very similar to the orb-shell cockles, but the shell is smaller and not so regular. There are about fifteen species of *Pisidium* in Britain, the largest being c. 10 mm in length.

Great Pond Snail *Lymnaea stagnalis*

The pond snail family (Lymnaeidae) contains a number of species found in the fresh waters of Europe, including Britain. Like most freshwater snails they do not, however, occur in very acid waters, such as moorland ponds and peat workings. The shape of the shell varies within a single species according to environmental factors, such as water temperature, water movement and food supply. This has led to the description of many local varieties as separate species. Pond snails glide over the bottom on a band of slime or mucus. The Great Pond Snail, in particular, can often be seen moving along under the water surface.

Pond snails belong among the pulmonates (Pulmonata) which breathe air by means of a lung and for this reason they have to come to the surface at intervals to renew their oxygen supply. If disturbed during this process they rapidly expel the air from the respiratory apparatus and fall like a stone to the bottom. Their mouth has a radula with thousands of tiny teeth with which they rasp algal growths off rocks and plants. Sometimes they bite off pieces of water plant. The Great Pond Snail, the largest member of the family, has a horn-coloured shell up to almost 60 mm long, with a characteristic pointed spire, which is almost as long as the mouth.

Wandering Snail *Lymnaea peregra*

This is a widely distributed pond snail in Europe, and probably the commonest species of *Lymnaea* in Britain. It is also possibly the most variable in the shape of the shell which usually measures anout 18 mm, sometimes a little more. It occurs in all kinds of standing and flowing water, and has even been recorded from brackish water. The spire of the thin shell is characteristically short but the mouth opening is much enlarged.

The related Dwarf Pond Snail (*L. truncatula*), which is only c. 10 mm long, serves as an intermediate host of the liver fluke, a flatworm parasite which infests sheep.

River Snail *Viviparus viviparus* (above, left)

Also known as the Freshwater Winkle, this is a gill-breathing snail widely distributed in Europe and in Britain found in slow-flowing waters as far north as Yorkshire. In addition to the comb-like gills this and related snails have a hard, horny structure, known as the operculum, which closes the shell opening when the snail withdraws its body. The shell is usually about 35 mm long, sometimes more, and is greenish-brown with three dark brown spiral bands (the dead shell illustrated is not typical). The last whorl is moderately enlarged. River Snails prefer waters with dense vegetation. The female produces about 50 eggs but these are not released to the outside world. Instead they develop within the body of the female until the embryos have completed their development. They are then set free into the water as young snails each about 10 mm long.

Ramshorn *Planorbis planorbis*

(above, right)

Great Ramshorn *Planorbarius corneus*

(below)

These two species belong to the family Planorbidae, of which there are several species in Europe, including Britain; they breathe by lungs. The Great Ramshorn has a tough, thick shell, up to 3 cm in diameter and red-brown. The whorls are cylindrical and five or six in number. This snail is fairly locally distributed in lakes, canals and slow-flowing rivers.

The Ramshorn is a somewhat smaller snail with a shell about 18 mm in diameter. The whorls are convex on one side, but flatter and keeled on the other.

The blood of ramshorn snails is red, due to the presence of haemoglobin which is an unusual feature in the molluscs. Aquarists often keep red specimens. These are so coloured because the shell has developed without pigment and the red colour of the blood is visible through the shell.

Medicinal Leech *Hirudo medicinalis*

(above)

This is one of several species of leech found in streams, ponds and lakes, usually in fairly shallow water among water plants. It attains a length of 10-15 cm. The young feed at first on invertebrates, later on newts and fishes. The adults also feed on the blood of mammals. This is now a rare animal in Europe, particularly in Britain. The much commoner Horse Leech (*Haemopis sanguisuga*), usually greenish or brownish and about the same length, does not feed on blood but preys mainly on snails, worms and tadpoles.

Water Mites Hydracarina

(below, left)

These are small, mainly freshwater animals with numerous species in all kinds of inland waters. The adults can be recognized by their unsegmented body and four pairs of legs. Depending upon their habits the legs of those living in flowing waters have hooks for attachment, those of lake forms have hairs for swimming. Water mites are all predators which feed mainly on small crustaceans and soft-skinned insect larvae.

Water Spider *Argyroneta aquatica*

(below, right)

This is the only spider species which lives completely underwater. It does, however, have to renew its supply of atmospheric oxygen. At the surface it extends the abdomen out of the water and takes air into special parts of the body where it is used for respiration. To increase the time spent underwater this spider builds a submerged web attached to plants, which it fills with air until it takes on the shape of a bell. The spider can then live in the bell, leaving it from time to time to hunt small crustaceans and the larvae of aquatic insects. For breeding the female lays 50-100 eggs at the top of a bell where they develop into young which are miniatures of the parents.

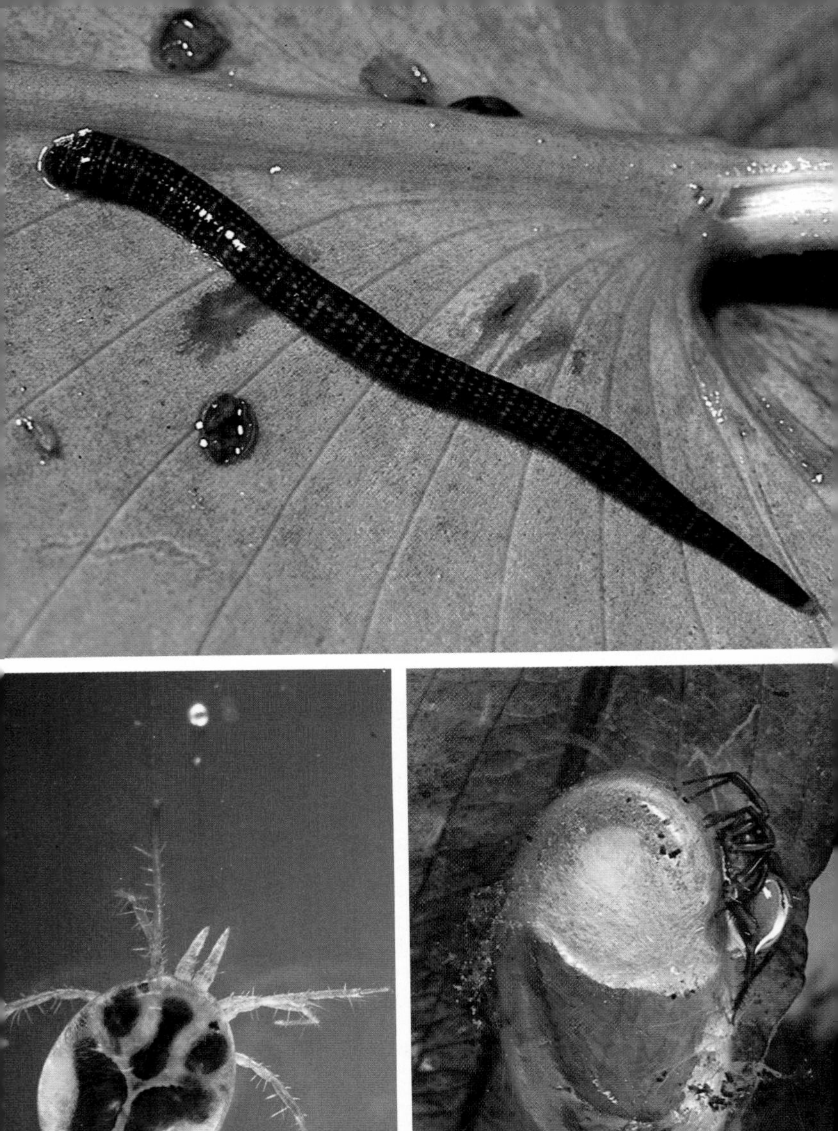

Water Slater *Asellus aquaticus*

Water slaters belong to the same group of crustaceans (Isopoda) as the wood-lice, which are so familiar, sometimes even in the house. *Asellus aquaticus* is common in fresh waters where the current is not too strong and where there is sufficient food in the form of detritus (mainly dead plant matter). It spends most of its time creeping along the bottom or climbing in among the plants, but it can also swim well. At mating time the two sexes swim together, the male gripping the female's body. The eggs are carried by the female on the underside of her body. The related whitish and blind *A. cavaticus* lives in subterranean waters and has been found in southern Britain.

Freshwater Shrimp *Gammarus* species

These are crustaceans belonging to the group known as amphipods (Amphipoda), most of which occur at various depths in the sea. The familiar sandhoppers on the seashore are amphipods. European fresh waters have several species belonging to the genus *Gammarus.* The males have a length of about 25 mm, the females slightly less. They occur in all types of fresh water, but mainly in streams and rivers where the content of oxygen is sufficiently high for their requirements. They also need a certain amount of calcium, which is incorporated in the carapace. Turning over a stone in a stream will nearly always reveal mayfly and stonefly larvae as well as some amphipods. With their laterally compressed body these small crustaceans are well adapted for slipping into narrow crevices. They feed primarily on plant and animal detritus. The male may carry the female about for as long as a week before mating. Depending upon her age the female can produce 20-100 eggs which she carries in a brood pouch under her body. These hatch into young which are miniatures of the parents, i.e. there is no larval stage. The juveniles moult several times before reaching sexual maturity. Species of *Gammarus* breed at all times of the year and they form an important food for fishes, particularly trout.

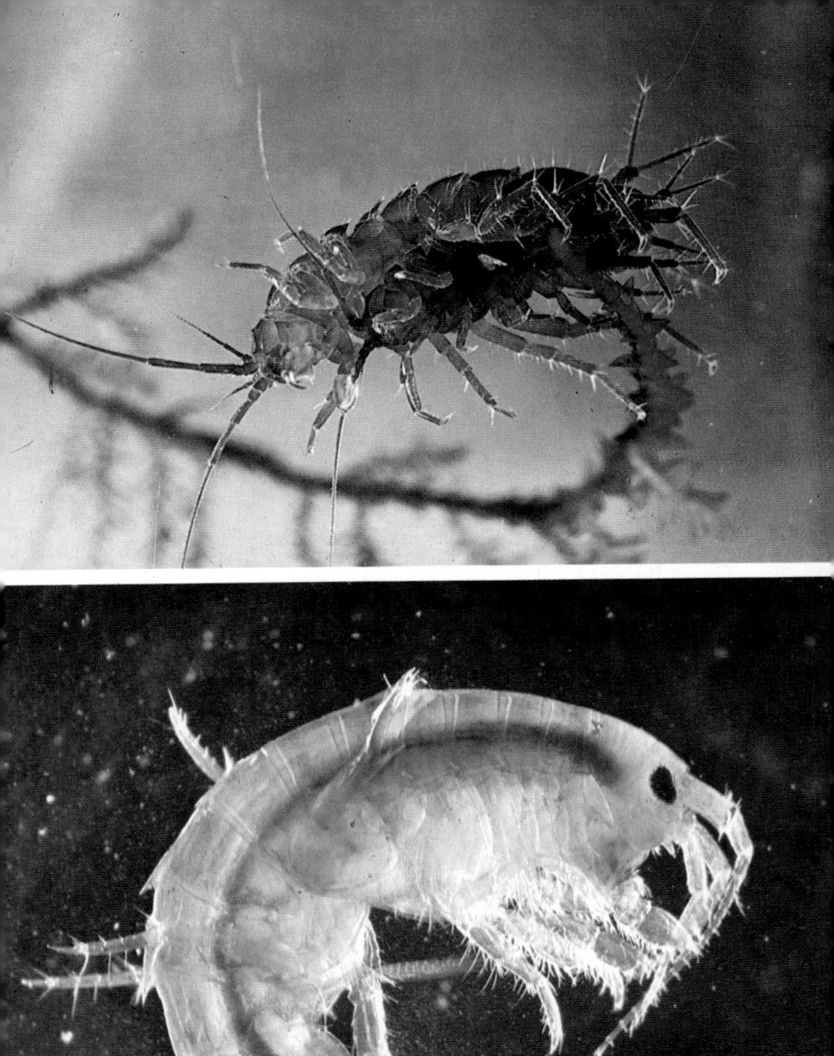

Freshwater Crayfish *Astacus* species

Most of the ten-footed crustaceans or Decapoda live in the sea, but there are few species in fresh waters. Those in Europe belong to the genus *Astacus.* They live mainly in rivers and streams with hard water that is rich in oxygen. They also require plenty of hiding-places under stones, in the shelter of overhanging banks or in among submerged tree roots. They remain concealed during the day but come out at night to search for food, moving about slowly on the thoracic legs. They feed on worms, aquatic insects, molluscs, sometimes fishes and amphibians and will not even refuse dead animals. The prey is first seized by the large pincers.

Species of *Astacus* pair in late autumn. usually in November. The male turns the female over and releases sperm on to her abdomen. The female then hides away and soon lays about a hundred pink eggs. These she attaches to her swimmerets (small limbs underneath the abdomen) where they are fertilized by the sperm left by the male. The eggs hatch in the following spring into tiny young which closely resemble the parents. The young cling to the female until their first moult and then start to live independently. In the early part of their life the juveniles moult several times a year, but after reaching sexual maturity in their 4th year they may only moult once a year. The new carapace becomes hard after about a week. During this period their skin is soft and they are particularly vulnerable. At this time they are also unable to feed and they remain hidden. On the continent of Europe the common species is *Astacus fluviatilis* in which the claws are reddish on the underside, but in Britain the true native species is *A. pallipes* which is smaller and has white claws. In some areas the populations have been much reduced by a fungal infection known as *Aphanomyces astaci.*

Water-fleas *Daphnia pulex* (above, left) and *Bosmina* species (above, right)

The water-fleas or Cladocera are small crustaceans, mostly 2-3 mm long, and almost exclusively restricted to fresh waters, although species of the genera *Podon* and *Evadne* occur on the seashore. There are some 60-80 species of Cladocera in Europe, mostly widely distributed and colonizing almost every type of standing water. Some species, such as *Daphnia pulex*, are cosmopolitan. Many live in the shallow waters of lakes with dense vegetation. Their hopping movements are caused by the sudden jerking of the two large, branched antennae. Under the microscope it is also possible to watch the beating of the paired limbs which are hidden within the two-valved shell. These serve primarily to waft in food, mainly planktonic algae, which is then filtered by a complicated apparatus before it reaches the alimentary tract. The eggs are usually carried in a dorsal brood pouch where they develop, without a larval stage, into small water-fleas. For much of the year these eggs hatch into females, a phenomenon known as parthenogenesis. When the environmental conditions become unfavourable, due to low temperatures, drought or lack of food, the female lays eggs some of which hatch into males, and she also lays "winter eggs" that have to be fertilized. These eggs are enclosed in a hard casing and they only hatch on return of favourable conditions to produce a new generation of parthenogenetic females, which will quickly build up a new water-flea population. Other genera with several species include *Moina, Ceriodaphnia* and *Chydorus*, while there are predatory forms in the genera *Polyphemus* and *Leptodora.*

Copepods *Cyclops* species (below)

The copepods (Copepoda) are small crustaceans, about 2 mm long, most of which live in the sea. *Cyclops* is one of the best-known freshwater genera. They have no shell and the body is divided into segments. The female carries the eggs at the back of the body and these hatch into larvae. unlike the eggs of water-fleas. Some copepods feed by filtering microscopic planktonic algae, others are predatory.

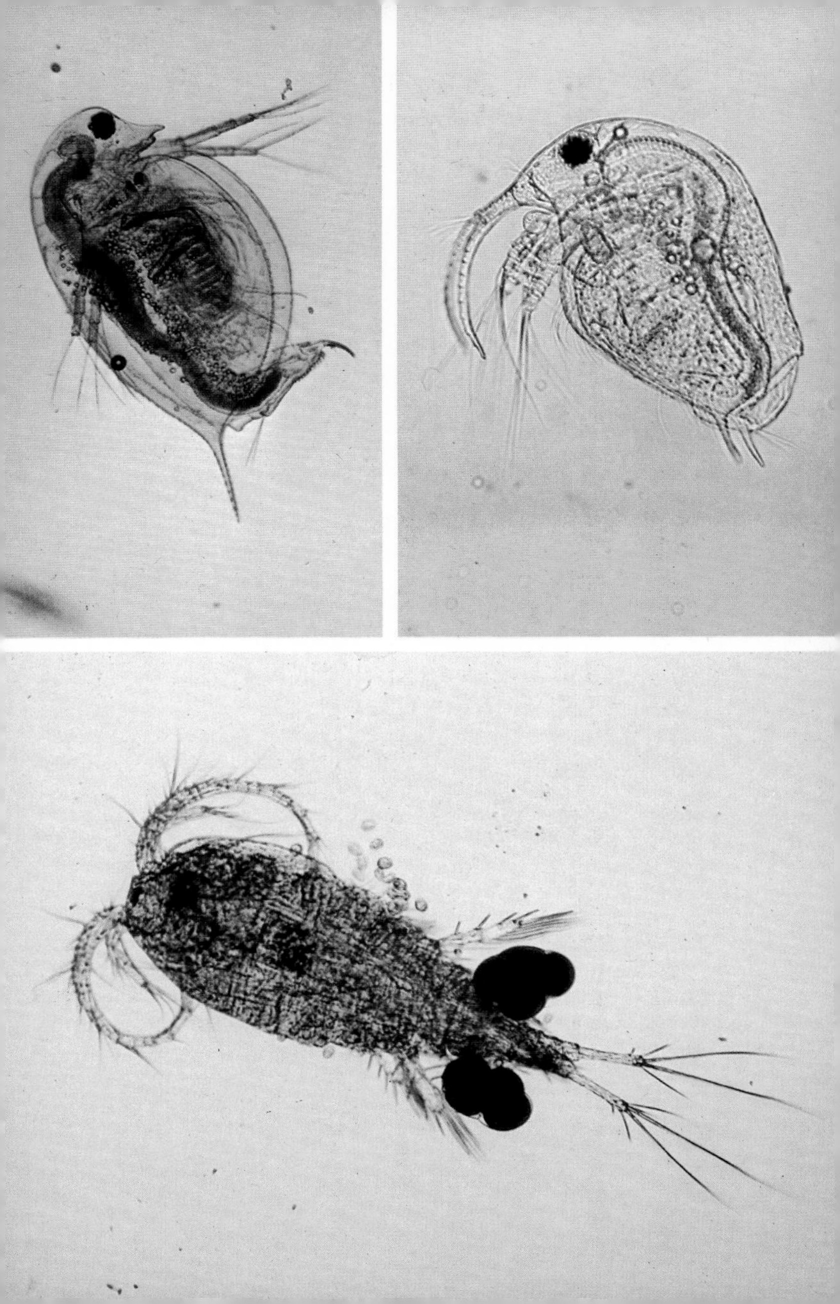

Mayflies Ephemeroptera

The mayflies form a distinct Order within the Class Insecta, and there are about 60 species in Europe (46 in Britain). They are easily recognized by the fact that, when at rest, the wings are held vertically above the abdomen. The front wings are triangular and relatively large, the hind wings small or absent. The antennae are very short and there are two or three long "tails" at the end of the abdomen.

Depending upon the species the sexually mature mayfly may only live for a few hours, or usually 2-3 days, although some live for up to three weeks. During the day they remain hidden among marginal vegetation, and start their courtship in the evening. Swarms of males dance up and down, and if a female approaches several males will fly towards her. The fastest seizes her below the body and mating takes place. They then separate before reaching the ground. The male dies shortly after this while the female lays her eggs before dying. A single female may lay between a hundred and some thousands of eggs. Some fly low over the water dipping the abdomen in as the eggs are released, others creep beneath the surface and lay eggs on stones or water plants.

In most species the aquatic larval period lasts about one year. Mayfly larvae, known as nymphs (below), are easily recognized by the tracheal gills arranged along each side of the abdomen and the three long "tails". The larvae of the genus *Epeorus* have only two "tails". Some larvae, such as those of *Ephemera*, move about in the bottom mud of slow-flowing streams. Others that live in fast-flowing waters have a flattened body and sometimes attachment organs. There are also swimming nymphs in standing waters, but most creep about on the substrate. As far as is known they all feed on algae and plant detritus.

Between the nymph and the adult or imago stage, mayflies have a winged sub-imago stage which may last from a few minutes to about 30 hours, depending upon the species. They only become sexually mature after the sub-imago has moulted into the imago.

Stoneflies Plecoptera

The Order Plecoptera has about 3000 species distributed throughout the world, of which some 125 occur in Europe, but only 34 in Britain. The larval stage is entirely aquatic, and almost all the larvae or nymphs (below) prefer running waters. In mountain streams the nymphs can be found on and under stones in the company of mayfly and caddisfly nymphs. They can be distinguished from the similar mayfly nymphs by having only two jointed "tails" or cerci at the end of the abdomen, and by lacking external tracheal gills on the abdomen.

Stonefly nymphs require water with a high content of oxygen, hence their preference for clear, fast-flowing streams. They are very sensitive to pollution. To prevent being swept away by the current they usually live under or on the lee side of stones, or in among water plants. There are three types of feeding. The first group, with small species (e.g. in *Nemoura*) feed on diatoms and plant detritus. In the second group with medium-sized forms (e.g. *Chloroperla*) the nymphs feed on invertebrates and plant matter. The third group with large forms (e.g. *Perla*) are among the most voracious of all the invertebrates living in running water. Smaller species mostly take in oxygen through the body surface but larger nymphs have tufts of gills on the body. The larval period may last from one to three years, the larger species taking longer than the smaller.

Adult or imago stoneflies are medium-sized, inconspicuously coloured insects which are always found in the vicinity of water. They can easily be recognized by the two cerci and by the roughly equal-sized membranous wings which are held flat over the abdomen when the insect is at rest. The identification of the different species is difficult and really a task for the specialist.

Pondskaters *Gerris* species

These are true bugs belonging to the group known as the Heteroptera. In Europe there are about 10 species of *Gerris* in the family Gerridae, some of which occur in Britain. They are widely distributed in quiet waters, where they usually live in groups. They have become adapted for living actually on the water surface, for they have unwettable legs which do not sink through the surface. Also the underside of the abdomen has hairs which are water-repellent. They skate about on the surface, hunting for insects and other small prey. The hind legs are used for steering, the middle legs for the actual locomotion, while the considerably shorter front legs serve in seizing the prey.

As in all the bugs, pondskaters undergo incomplete metamorphosis. This means that the eggs hatch into larvae which, after several moults, develop into the adult or imago. A pupal stage is lacking. Some pondskater species have two, others only one generation in the year, so it is quite possible to find both larvae and adults living alongside one another.

Water Cricket *Velia currens*

Like the pondskaters, the water crickets in the family Veliidae live on the surface, but in this case primarily of running waters and usually close to the banks. They can run extraordinarily fast, even against the stream, and can also dive below the surface. They feed mainly on small invertebrates that land on the water. In the species of *Velia* mating takes place in the spring and there are five larval stages.

The adults are 6-8 mm long and they usually have no wings. They are more stoutly built than the pondskaters and their middle and hind legs are shorter. They catch their prey on the snout-like rostrum, not with the legs. Several species occur in Britain.

Common Water-boatman
Notonecta glauca (above)

This is one of the bugs known as backswimmers (family Notonectidae). There are four species in British waters. They can be seen at most times of the year at the surface of standing waters. They characteristically swim in an upside-down position and hang beneath the surface to take in air which can be held among water-repellent hairs on the underside of the body. This air provides buoyancy and is responsible for the remarkable inverted position of the insect. The hind legs, which have fringes of hairs, are used in swimming. Backswimmers are voracious predators which seize invertebrates with the front legs and suck the body contents with the beak or rostrum.

Water Stick-insect *Ranatra linearis*
(below, left)

Water Scorpion *Nepa cinerea*
(below, right)

These two aquatic bugs belong to the family Nepidae, and both have a preference for standing or slow-flowing waters, where they live close to the shore, usually among plants. They are common in Britain. They are not good swimmers, but lie in wait for their prey, either among plants or on the bottom. The long "tail" is actually a respiratory tube which the insect pushes up through the water to take in air. Passing prey is seized by the front legs and conveyed to the head where its body contents are sucked in by the rostrum. *Ranatra* feeds on water-fleas and other small invertebrates, while *Nepa* can take larger prey, including tadpoles and small fishes. Excluding the respiratory tube *Ranatra* is 4 cm long, *Nepa* 2 cm. The eggs hatch in May-July and the larvae become adult, usually in September, after undergoing five moults. Both species spend the winter in the imago or adult stage.

Great Diving Beetle *Dytiscus marginalis*

There are several beetles that are adapted for aquatic life, and many of their larvae also live in the water. The Great Diving Beetle is a very typical and striking example and it is easily recognized. It belongs to the family Dytiscidae. The adults are 3-4 cm long and are dark olive-green above, yellowish-red below. The flat, oval body is outlined by a broad yellow band. The sexes are easily distinguished, for the male has smooth elytra (or wing-covers) and suckers on the front legs with which it grips the female during mating. The female, on the other hand, has furrowed elytra (below). Naturally, both sexes have legs adapted for swimming, in particular the hind legs which are broad and hairy. Water bugs have similar limbs.

The Great Diving Beetle is a widely distributed insect throughout Europe, including Britain, mostly living in ponds and lakes. If one of these dries out the beetle flies to the next suitable body of water. Both the adults and the 6 cm long larvae (above) are voracious predators which take any aquatic animal of a suitable size, including tadpoles and small fishes. The adults use their mouthparts to eat their prey, cutting it up into fragments before swallowing it. The larvae, on the other hand, use their sharp, pointed mandibles to seize and pierce prey. Each mandible has a thin canal through which a digestive secretion is pumped into the prey where it digests the contents. The larva then sucks out the liquefied contents. It has been reckoned that during its larval period of six weeks, each larva moults three times and eats about 20 tadpoles per day. The larva creeps out on to land, buries itself in the soil and becomes a pupa. After a further two weeks the adult beetle emerges. Both larva and adult require atmospheric oxygen. The beetle rises to the surface at intervals to collect air under its elytra before diving down again. The larva also comes to the surface but takes air in at two pores at the rear end.

Fly and midge larvae Diptera

A very large number of flies of the Order Diptera have aquatic larvae. The Order can be divided into two distinct Sub-orders, the Nematocera (e.g. mosquitoes, midges, craneflies) which mostly have a slender body and slender antennae, and the more stoutly built Brachycera with short antennae. They all have only one pair of functional wings, the hind pair being reduced to tiny balancers or halteres.

Fly larvae vary considerably in form depending upon environment and habits. The phantom larva of *Chaoborus*, formerly *Corethra* (above), which lives in lakes, is so transparent that all the organs of the body can be seen through the skin. It has no respiratory apertures, but takes in oxygen through the skin. Its antennae are modified for catching prey, mainly small crustaceans. Phantom larvae float horizontally in the water at various depths depending upon the amount of air in the front and rear air sacs, which are easy to see on account of their dark pigmentation. The pupae hang vertically in the water and rise to the surface shortly before the imago or adult fly emerges.

The larvae of blackflies of the family Simuliidae (below, left) live almost exclusively in running water. They attach a layer of mucus to a stone and then fasten themselves to this by means of hooks at the rear end of the body. These larvae sieve food particles (algae and detritus) out of the water, using two groups of bristles on the head.

The larvae of soldierflies of the family Stratiomyidae (below, right) live in springs, ponds and lakes. They breathe air at the surface through two pores or spiracles at the rear end of the body.

Finally, the true or non-biting midges of the family Chironomidae have worm-like larvae which live in the bottom and form an important part of the diet of many fishes.

Damselfly larvae *Calopteryx* species

(above)

Demoiselle Agrion *Calopteryx splendens*

(below)

The Order Odonata contains two quite distinct groups, the Sub-order Zygoptera or damselflies and the Sub-order Anisoptera or true dragonflies. The damselflies have two pairs of similarly shaped wings which are held vertically over the abdomen when the insect is at rest. Their eyes are widely separated and set on the sides of the head. Damselflies have a slow, fluttering flight and they often sit on shore plants waiting for prey, which consists of flies, moths and wasps. The true dragonflies are more stoutly built, with hind wings that are broader than the front wings and large eyes that meet or almost meet on top of the head. The wings are held horizontally when the insect is at rest. Dragonflies have very strong and rapid flight. The legs can be held in the form of a basket in which the insect catches prey in flight. The insects caught in this way are either eaten in the air or taken to a perch for consumption.

The Demoiselle Agrion (below) is one of the most striking of the damselflies and is widely distributed in Europe, including Britain, western Asia and northern Africa, occurring at altitudes up to 1200 m. The body is 5 cm in length and wing span is 7 cm. The female is not so brightly coloured as the male. They fly rather slowly over the water when the sun is shining.

The larvae or nymphs of damselflies are easy to recognize as they have a relatively slender body with long external gills at the rear end.

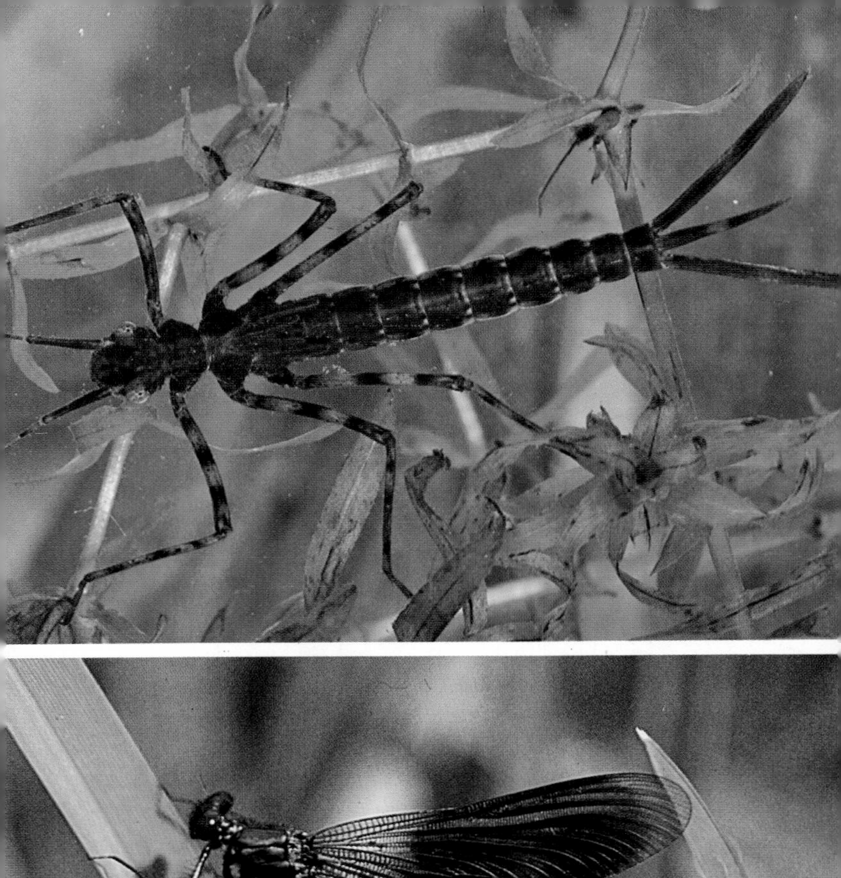

Dragonfly and damselfly larvae
Odonata

On any collecting trip to a pond or lake it should be possible, with the help of a suitable net with a long handle, to fish up dragonfly nymphs from the bottom. As in the case of the adults it is quite easy to distinguish the nymphs of damselflies from those of the dragonflies.

Damselfly nymphs have an elongated slender body. At the end of the abdomen there are three conspicuous tracheal gills, structures which do not occur in this form in any other aquatic larvae.

On the other hand the nymphs of true dragonflies are more powerfully built with a stout body, and without external gills.

Both types of nymph have two wing buds on the upperside behind the head and, most characteristically, a structure known as the mask below the head. The mask is folded back when at rest but can be rapidly shot out to catch prey, as in the upper photograph, where a mosquito larva is being seized. These nymphs are all predators which catch small crustaceans, worms and aquatic insects, sometimes even water slaters, tadpoles and small fishes. The upper photograph on p. 77 also shows a damselfly nymph, but with the mask folded back. The lower photograph on p. 79 shows the nymph of a true dragonfly of the family Libellulidae.

Dragonfly nymphs have numerous tracheal gills on the inside of the hind gut, and these serve in respiration. Water is pumped rhythmically in and out of the anus. If water is ejected very rapidly the nymph will be shot forward like a rocket, an action which may help it to escape an enemy.

Black-lined Orthetrum
Orthetrum cancellatum

A widely distributed dragonfly throughout Europe, including Britain. It is mostly seen near and over lakes where it flies up and down along the shores and over neighbouring damp meadows. This species reaches a body length of 5 cm and the wing span is up to 9 cm. It is active between May and September. The broad abdomen is characteristic. In males this is blue with dark grey on the last segments, but in females it is dark with yellow longitudinal stripes. This species might be confused with *Libellula depressa* but the latter has dark brown wing bases. The nymph of *O. cancellatum* lives among plants on the bottom and hunts for worms, the larvae of aquatic insects and other invertebrates. Depending upon the water temperature and the supply of food, larval development may take several years. The adults feed on insects which they catch in flight.

Southern Aeshna *Aeshna cyanea*

This large dragonfly occurs in most parts of Europe, eastwards into western Asia and in parts of northern Europe. It is not uncommon in Britain where it can be seen flying near or over lakes and canals from the middle of June to the beginning of November. The body is c. 8 cm long and the wing span 11 cm. The pattern of the abdomen is an important character for the identification of the different species. The photograph shows a male. Old females have a dark red-brown abdomen with pale green areas. The eggs remain dormant over the winter and hatch at the end of April or beginning of May. The nymph usually takes about two years to reach the adult stage.

Alderflies *Sialis* species (above, left)

Alderflies live in the vicinity of water, mostly along the banks of rivers and lakes. At rest the wings are held in the form of a roof. They differ from caddisflies in having wings that are almost equal in size. Alderfly larvae live as predators on the bottom.

Caddisflies Trichoptera

Members of the Order Trichoptera are widely distributed in Europe, including Britain. The adults (above, right) are not difficult to recognize. When at rest the wings are held over the body in the form of a roof. This posture is also seen in some moths, but the latter always have scales on the wings, where caddisflies only have hairs. In addition, they do not have the sucking proboscis characteristic of moths. In almost all caddisfly species the adults are inconspicuously coloured. Brown and grey tones predominate, sometimes with a few markings. When at rest the long antennae are stretched out forwards. Caddisflies are mostly seen close to water between the beginning of June and the end of August. They usually live for 3-4 weeks.

Caddisfly larvae live in water and most of them build a case in which they live. A few, such as the larvae in the genera *Hydropsyche* and *Rhyacophila*, do not build a case. The case is constructed by the front limbs and the mouthparts. Depending upon the species the case may be made of fragments of leaf or reed, tiny twigs (below, right), or other plant material, but some species use sand grains or small pieces of gravel. The case protects the larva from some predators. Most larvae creep about in search of food, taking plant fragments and invertebrates. In running waters there are net-building species which build nets in which they catch food particles (below, left). Caddis larvae pupate for two to three weeks and then emerge as adults which have to climb out of the water.

Brown Trout *Salmo trutta fario*

A widely distributed fish over the whole of Europe, eastwards into western Asia, and southwards to the Black Sea, in rivers with cold water rich in oxygen. Owing to their well-flavoured flesh Brown Trout are extensively reared in trout farms. This is the stationary form of *Salmo trutta*. The migratory form is the Sea Trout (*S. trutta trutta*) which lives in the sea but ascends rivers to spawn. Brown trout have a torpedo-shaped body which is usually greenish or brownish with a whitish belly. The upperparts are marked with round, black spots and along the lateral line there are red spots with pale edges. Spawning takes place in late autumn and winter. The female beats her tail to make a pit or redd in the gravel, in which she lays about 1000 eggs. After fertilization by the male these are covered over with gravel. Older individuals establish territories which they defend against intruders. Trout feed mainly on small crustaceans and aquatic insects, sometimes even small fishes.

Grayling *Thymallus thymallus*

Found in many parts of Europe, but not in the Iberian Peninsula, Italy, Scotland, large areas of Scandinavia or the Balkans. The general distribution is rather scattered because this fish has rather special environmental requirements. Grayling live in clean, fast-flowing waters rich in oxygen with a firm substrate. In certain areas their numbers have been adversely affected by pollution. The species is easily recognized by its elongated body, the presence of an adipose fin and the characteristic long dorsal fin. The body assumes a reddish tint at spawning time which is from March to June. The female beats out a spawning pit with her tail in which she lays the eggs. After these have been fertilized by the male they are covered with gravel, and they hatch in about two weeks. Grayling become sexually mature at 3-4 years and may grow to a length of up to 60 cm. They feed on worms, snails, insect larvae, fish eggs and small fishes.

Whitefish *Coregonus* species

The genus *Coregonus* contains migratory forms living in the coastal waters of the North and Baltic Seas, and stationary forms in large and deep lakes in the Alps, northern Germany and northern Europe, including Britain. They are not at all easy to identify with any degree of accuracy. The common characteristics include the elongated body, which is laterally compressed, and the very pointed head. There is also a small adipose fin between the dorsal and caudal fins, the latter being deeply forked. Whitefish spawn in late autumn and winter. Some forms lay their eggs in the open water, others spawn close to the shore over sand or gravel. Forms living in open water feed primarily on zooplankton animals, bottom-living forms on insect larvae, worms and other invertebrates.

Arctic Charr *Salvelinus alpinus*

This is a very widely distributed member of the Salmon family with populations scattered throughout Europe, temperate Asia and North America, in deep cold lakes. There are also migratory forms in the coastal waters and rivers of the Arctic. In Europe there are Charr in lakes in Scandinavia, the Alps, Scotland and Ireland, in some places up to altitudes of 2000 m. The most prominent features are the bright white edges to the pectoral, ventral and anul fins. At spawning time, which is generally between September and January, the belly becomes reddish, whereas at other times it is yellowish-white. As in the Salmon, Brown Trout and the species of *Coregonus*, Arctic Charr have a typical adipose fin. They reach a length of 25-40 cm. There are numerous varieties, some of which feed on planktonic crustaceans, while others prey on bottom-living invertebrates or fishes.

Roach *Rutilus rutilus*

A common fish in standing and slow-flowing waters from the Pyrenees to Siberia, but not found in Scotland, Norway, Italy or the western Balkans. Roach live mainly in shoals fairly close to the banks and feed on a varied diet, ranging from aquatic invertebrates to vegetation. They spawn in April-May when they may congregate in large shoals. The eggs are attached to underwater vegetation and the larvae are at first also attached. A female may lay over 50 000 eggs which hatch in 4-10 days. Roach mostly become sexually mature at three years. The body is laterally compressed and the front end of the dorsal fin lies above or just behind the front of the ventral fins. The iris is red. Adults may reach a length of up to 50 cm. but are usually 25-30 cm. This is a very variable species and several subspecies have been described. Some live in brackish water but move into the rivers to spawn.

Rudd *Scardinius erythrophthalmus*

A widely distributed fish in standing and slow-flowing waters in western, central and eastern Europe, but absent from Scotland, the Iberian Peninsula and large areas of Scandinavia. They live mainly among vegetation close to the banks. The most characteristic feature is the reddish tint of the fins. The eyes are yellowish-red or yellow. The body is usually yellowish-green at the rear, more silvery in front and the belly is whitish. Rudd are omnivorous, feeding mainly on plants such as milfoils and pondweeds, but they also take some invertebrates. They spawn between April and June and the eggs and early larvae adhere to plants and rocks. The larvae feed on planktonic crustaceans. The species reaches sexual maturity at an age of 2-3 years. It is of no commercial importance, but is often caught by anglers.

Barbel *Barbus barbus*

A bottom-living member of the family Cyprinidae found from the Iberian Peninsula eastwards through central Europe to the Black Sea, occurring in some parts of England but not in Scotland or Ireland. Barbel require clear running waters rich in oxygen with a gravel or sand substrate. They have an elongated, slender body, a down-facing mouth and four barbels on the upper lip. They remain hidden by day, emerging at night to hunt for food, which consists primarily of worms, molluscs and insect larvae. Barbel reach a maximum length of 90 cm, and a weight of 8.5 kg. Such large individuals also take small fishes. They spawn between May and July, moving upstream in large shoals. A female lays 3000-9000 yellow eggs which at first adhere to stones. They hatch in 10-15 days. Barbel become sexually mature at the end of their third or fourth year.

Common Bream *Abramis brama*

Widely distributed in Europe from the Pyrenees eastwards to the Urals, but not south of the Alps. Found in eastern England, southern Scotland and parts of Ireland. Common Bream prefer large eutrophic lakes, but also occur in the lower, slow-flowing reaches of rivers. The body is high-backed and laterally compressed. They mostly grow to a length of 30-50 cm, but individuals up to 75 cm, weighing up to 6 kg, have been recorded. As would be expected for a bottom-living fish they feed on worms, molluscs, small crustaceans and insect larvae. They spawn in May to June, gathering in large shoals in shallow water. A female may lay c. 300 000 eggs which hatch in 3-12 days. The larvae at first remain attached to aquatic plants. Common Bream become sexually mature at an age of 3-4 years.

Tench *Tinca tinca*

A widely distributed fish throughout Europe and temperate Asia, including England, but not extending into the far north. It prefers warm waters with dense vegetation. The main distinguishing features are the brownish-green coloration, the rounded fins and the thick slimy skin with deeply embedded scales. In some areas Tench are cultivated in artificial ponds. A golden form is often exhibited in public aquaria.

Carp *Cyprinus carpio*

Widely distributed throughout Europe and Asia, except in the more northerly areas. Introduced into England, Wales and Ireland many centuries ago, but it does not occur in northern Scotland. It is extensively farmed and is the most important commercial freshwater fish in Europe. Carp prefer deep, standing or slow-flowing waters with a muddy or sandy bottom and dense vegetation, usually spending the day on the bottom. At night they become active, searching for bottom-living invertebrates, particularly insect larvae and small crustaceans. Large Carp also eat newts and small fishes. They may grow to a length of 120 cm, and weigh 25-30 kg. The original form is the scaled Carp with a complete covering of scales, but there are also mirror Carp (opposite, below) with a small number of large, mirror-like scales and leather Carp with few or no scales. Transitional stages between these three main forms are not uncommon. Carp normally spawn between May and July, but commercial breeders can control this with the help of hormones, so that the fish come into breeding condition in a short time. A female Carp may spawn 120 000-150 000 eggs per kg weight, sometimes more. These adhere to plants and hatch in 3-5 days. Under natural conditions most of the larvae and juveniles will perish, but hatcheries are able to rear large numbers of young which can be transplanted into suitable waters at an age when they have a better chance of survival.

Bitterling *Rhodeus sericeus amarus*

This small member of the Cyprinidae occurs in Europe north of the Pyrenees and Alps eastwards into western Asia, but does not occur in Ireland, Denmark and Scandinavia. It is not native to Britain but has been introduced in a few places. It is a gregarious fish that lives in shallow water close to the banks of slow-flowing rivers and of lakes, where it feeds on worms, small crustaceans and aquatic insects as well as on plants. In April to June the female develops an ovipositor which is several cm long (opposite, above, lower fish). She inserts this structure between the two valves of a Swan Mussel and lays her eggs in its gills. The male sheds sperm over the mollusc and this is drawn in with the respiratory current to fertilize the eggs. The latter hatch in 2-3 weeks and the larvae remain in the mollusc's gill chamber until they have consumed most of the contents of their yolk sac. They then pass out through the mollusc's exhalant siphon.

Three-spined Stickleback
Gasterosteus aculeatus

A common fish in the cold and temperate fresh and coastal waters of Europe, Asia and North America. The three separate spines on the back can be erected, thus forming a protection against being swallowed. The species, which reaches a length of up to 8 cm, occurs in a migratory marine form which in spring moves from the sea coasts into fresh water, and a stationary form which spends its whole life in fresh water. Juveniles of the migratory form grow up in fresh water and then move into the sea. In doing so they have to experience considerable fluctuations in salinity. At the start of the spawning period the male develops a bright red belly and establishes a territory in which he builds a best of plant fragments. He then entices a female into the nest where she lays her eggs. He then sheds sperm to fertilize them. This female is then driven away and he may repeat the process with other females. Finally, the male guards the eggs and later the fry for about a week.

Perch *Perca fluviatilis*

Distributed throughout Europe and parts of Asia, with the exception of the Iberian Peninsula, most of Italy, northern Scotland, western Norway and the western Balkans. Perch (family Percidae) live mainly in standing and flowing waters at altitudes up to c. 1000 m, preferring clear water and a hard substrate. They reach a length of up to 45 cm and are eaten in some places. Their main distinguishing features include the spiny first dorsal fin, the conspicuous transverse bars on the flanks and the reddish ventral and anal fins. They spawn between March and June, the eggs being laid in long, gelatinous strands among vegetation in shallow water. They hatch in 2-3 weeks. Perch mostly become sexually mature when they are three or four years old. They have a very varied diet. Younger individuals feed mainly on small crustaceans in the plankton or on worms, midge and other aquatic larvae on the bottom, while the larger individuals take small fishes.

Miller's Thumb *Cottus gobio*

Also known as the Bullhead, this species is found throughout Europe but is absent from Ireland, most of Scotland and northern Scandinavia. This is a non-migratory fish which prefers clear, fast-flowing streams and small rivers, with a gravelly bottom, and also occurs in alpine lakes at altitudes up to 2200 m. Miller's Thumbs live on or near the bottom. In keeping with this the body is wedge-shaped and the head broad and flat. They mostly grow to a length of 10-15 cm. By day they live hidden among stones and tree roots, and emerge at twilight to feed on small bottom-living invertebrates, fish eggs and fry. Spawning takes place between February and May. The male makes a shallow spawning pit, often under a stone, in which the female lays her eggs. These are guarded by the male and, depending upon the water temperature, hatch in 3-6 weeks. The young grow rapidly and are sexually mature at the end of the second year.

Pike *Esox lucius*

Distributed throughout Europe (except southern Italy), temperate Asia and North America; introduced into Ireland and Spain many centuries ago. This is a non-migratory freshwater fish living in running or standing waters, particularly in shallow areas near the banks. There it can lie in wait to seize prey with the sharp teeth. It feeds on invertebrates, amphibians, water birds and particularly on other fishes. Pike spawn between February and May among reeds and other plants in shallow water, the female laying 15 000-20 000 eggs per kg body weight which adhere to plants and hatch in 2-3 weeks. The fry feed at first on small crustaceans. The young are very voracious and may reach a length of 15-30 cm in one year, becoming sexually mature in the second year.

Wels *Silurus glanis*

A large catfish distributed in eastern Europe eastwards into temperate Asia. They are not native to Britain but have been introduced into certain private lakes, e.g. at Woburn Abbey in Bedfordshire. Wels prefer to live in warm lakes and large rivers with a soft bottom where they lie concealed by day in cavities and deep pits. They become active at night. They spend the winter singly or in small groups without feeding. The body is slimy with a broad, flat head, a very short dorsal fin and a very long anal fin. The broad mouth has two long barbels on the upper jaw, four short ones on the lower jaw. Wels usually grow to a length of 100 cm, some to 300-400 cm, and they feed on other fishes, frogs, water birds and small mammals. They spawn between May and July and, depending upon the temperature, the eggs hatch in 3-10 days, giving fry that are 7 mm long. They become sexually mature in their third to fifth year.

Alpine Newt *Triturus alpestris*

A widely distributed newt ranging from northern Spain, the Alps, northern Italy and Greece to the Carpathians, but not found in Britain. These attractive amphibians can be seen as early as February in both standing and running waters in hilly country and particularly in mountains up to altitudes of 3000 m. Spawning may continue until May when the adults leave the water and go on land. It is interesting that the larvae spend the following winter in the water because, due to the high altitude, they are unable to complete their development within a single summer. During the breeding period the males have a low, smooth-edged dorsal crest which is continuous with the tail crest. The most striking feature is, however, the orange-red belly which is usually unmarked. These striking underparts are separated from the slate-grey upperparts by numerous dark spots. Alpine Newts grow to a length of c. 11 cm.

Smooth Newt *Triturus vulgaris*

A common newt throughout most of Europe (including Britain) and western Asia, but it does not occur in southern France, the Iberian Peninsula, southern Italy and certain of the islands in the Mediterranean. It is perhaps the most undemanding of the European newts, appearing in ponds and ditches from early spring onwards. It has been recorded in the mountains up to altitudes of c. 1000 m. The body is rather slender and c. 10 cm long. The breeding season is usually in April and May and at this time the males develop a tall wavy dorsal crest which is not so toothed as in the Crested Newt and which is continuous with the tail crest. The head has five longitudinal stripes, a quite characteristic feature of this species. The crests disappear after the breeding season and Smooth Newts then lead a secretive, nocturnal, terrestrial life until the following spring. Like other newts they feed on small invertebrates.

Crested Newt *Triturus cristatus*

Widely distributed in Europe north of the Alps, from Britain and central France in the west to central Russia in the east, and central Sweden to the north. Within this wide range Crested Newts occur in canals, ponds and lakes in low-lying and hilly country up to altitudes of c. 1000 m. They are not as common as they used to be, due largely to water pollution and land drainage for agriculture. They grow to a length of 15-16 cm. In the breeding season, between March and July, the males are easily distinguished by their tall, toothed dorsal crest which is clearly separated by a deep indentation from the tail crest. Females have no dorsal crest, but only a fold of skin above and below the tail. After the breeding season both sexes leave the water and the crests and skin folds disappear.

Palmate Newt *Triturus helveticus*

This is a West European species, with a range extending from the Iberian Peninsula through France, Britain, Belgium, the Netherlands to western Germany and Switzerland, usually preferring wooded, hilly country. In the breeding season, from the middle of March onwards, they can be found in ditches and streams, sometimes also in standing waters. At this time the males have a quite characteristic thin filament 1 cm long at the end of the tail and a low smooth-edged crest on the back and tail. By early summer they are moving out on to land where they spend the winter. At this time the tail filament and crests disappear. The female has neither tail filament nor crest. Palmate Newts usually have brownish-olive upperparts, often with small dark spots, and usually a dark, longitudinal stripe running through each eye. The central part of the belly is bright yellowish with white on each side and usually without spots.

Edible Frog *Rana esculenta*

Distributed from France throughout the whole of central Europe to Italy, Denmark, southern Sweden and eastwards to Romania, Hungary and western Russia. There are a few small colonies in England, but these may all have been introduced. Edible Frogs live in small or large bodies of water with dense vegetation, particularly along the banks. They grow to a length of 7-10 cm, exceptionally to 12 cm, and they are easy to recognise by the grass-green coloration with yellow and dark brown marbling on the hind legs. They often have a pale, usually yellowish-green, longitudinal stripe on the back, but never the dark, wedge-shaped marking behind each eye which is typical of the Common Frog. In the breeding season in May they emit a loud, raucous call. In doing so the whitish vocal sacs protrude on each side behind the angle of the mouth. In contrast to Common and Green Toads, Edible Frogs lay their eggs in large clumps in the water, each containing several hundred eggs. A single female may lay a total of 5000-10 000 eggs. After 7-10 days these hatch into tadpoles which will metamorphose into frogs after about four months. The young frogs usually become sexually mature when they are three years old. The life span is about ten years.

Common Frog *Rana temporaria*

Found throughout Europe, including Britain (but not in Iceland, Ireland and northern Russia) and eastwards to temperate Asia and Japan. They spend the spring in the water, preferring small ponds and lakes, and the remainder of the year in swamps, damp meadows and parks, often at quite a distance from water. In the breeding season, in February-April, the males have a purring croak, made with the help of the two vocal sacs on the underside of the head. The eggs are laid in large clumps and the tadpoles metamorphose in about three months.

Common European Toad *Bufo bufo*

Widely distributed throughout Europe, including Britain, but excluding Ireland, Sardinia, Corsica, Malta, Crete and the Balearics. The range extends eastwards through temperate Asia to Japan. Common Toads grow to a length of c. 13 cm, the males being smaller than the females. They live mainly in woodland and gardens, but can also be found in meadows and even damp cellars. They are very benificial animals as they consume large numbers of slugs and insect pests. They enter the water for breeding in March-April. Mating takes place in the water, with the male grasping the female from behind. She lays 5000-7000 eggs in 2-4 rows in a band that may be 3-5 m long. This is in contrast to the position in the Common Frog where the eggs are laid in irregular clumps. The eggs hatch in about a fortnight and the tadpoles metamorphose in June when they are c. 3 cm long. Common Toads live for up to 30 years, occasionally even longer.

Green Toad *Bufo viridis*

Distributed in central, southern and parts of western Europe, but not in Britain, large areas of France, the Iberean Peninsula, Belgium and Holland. The range extends northwards to Denmark and southern Sweden. This is a mainly nocturnal toad, usually associated with quite dry, sandy areas in low-lying country. It has a shrill, high-pitched call which can be heard at night during the breeding period. Mating takes place in the water between the beginning of April and the beginning of June. The female lays the eggs in long strands, with a total of 10 000-12 000. The adults grow to a length of up to 9 cm. They can scarcely be confused with any other European toad as the dark green pattern with red dots on a pale background is quite characteristic.

Yellow-bellied Toad *Bombina variegata*

This colourful small toad is distributed throughout most of central and southern Europe, but does not occur in Britain, the Iberian Peninsula and southern Greece. It grows to a length of c. 5 cm. Although it can be found in the lowlands it is more typical of hilly country, sometimes up to altitudes of 1800 m. It lives in the shallow water of ponds and ditches, or in slow-flowing streams, preferring clear water with plenty of vegetation. The upperside is olive to grey-brown, sometimes with dark markings, the underside is bluish-grey to almost black with large lemon-yellow markings and white dots. This is typical warning coloration. When on land and threatened these toads turn over on to their back and expose the strikingly coloured belly to the intruder. The normal position is resumed when danger has passed. Yellow-bellied Toads dive very rapidly and bury themselves in the mud at the bottom. They also have a chemical method of defence in their poisonous skin secretions. These substances produce unpleasant inflammation of the eyes, nose and lips in man. Breeding takes place between the second half of April and August. The eggs are laid in small clumps on rocks and aquatic plants. They hatch in about a week.

The very similar Fire-bellied Toad (*Bombina bombina*) is found in low-lying country (up to 250 m) in eastern Europe, extending north to Denmark and southern Sweden, eastwards to the Urals and south to Bulgaria and Yugoslavia. The belly has a pattern of red blotches. It has the same behaviour pattern when threatened as the preceding species, and its breeding habits are similar. The young take about three years to reach sexual maturity. The life span is 20-30 years.

European Pond-tortoise
Emys orbicularis

Distributed in southern and central Europe, north-west Africa, extending into western Asia, but not found in Britain. In many areas it is becoming rather scarce. This is an amphibious reptile that lives in standing and slow-flowing waters that contain sufficient animal life. The diet consists of snails, worms, insects, small fishes, amphibians and some plant food. The prey is always eaten in the water. The carapace, which is usually up to 20 cm long, sometimes to 30 cm, has characteristic markings, which clearly distinguish this species from any of the land tortoises. Pond-tortoises are mainly active by day and in twilight and they like to lie out on rocks or banks to sun themselves. They quickly take to the water if disturbed. They swim well, using their webbed feet. Mating takes place in the spring. The female uses her tail to dig a pit in which she lays 3-16 eggs. These hatch after about two months and the young are then completely independent and start to feed in the water. The life span is up to 70 years.

Grass snake *Natrix natrix*

A common reptile throughout Europe up to about latitude 67°N in Scandinavia, but absent from Ireland and some of the Mediterranean islands. The range also extends to north-west Africa and western Asia. Grass Snakes usually reach a length of 120 cm, although specimens up to 200 cm have been recorded. They are normally associated with water, although not directly tied to it, and they swim and dive well. They feed on lizards, newts, frogs and fishes. They do not have venomous fangs and the prey is swallowed whole and live. The yellow, orange or white patch on each side behind the head is quite characteristic. When disturbed they hiss, raise the head and release an evil-smelling secretion from the anal glands. They seldom bite. Mating takes place in spring or early summer and the female lays 10-40 eggs in July, usually under stones or leaves. These hatch in 7-10 weeks and the young are then c. 20 cm long. The life span is up to 25 years.

Great Crested Grebe *Podiceps cristatus*

A common breeding bird throughout Europe, with the exception of northern Scandinavia, and living mostly on lakes. This is the largest of the European grebes. The others include the Red-necked Grebe (*Podiceps griseigena*), the Black-necked Grebe (*P. nigricollis*), the Slavonian Grebe (*P. auritus*) and the Little Grebe (see below). The Slavonian Grebe is a more northern species, breeding in Scandinavia and northern Scotland. The Great Crested Grebe is about the size of a Mallard and in the breeding season easily recognized by the large crest and the reddish-brown to blackish frill on the sides of the head. The courtship ceremony in April-May is particularly attractive. The nest is a large floating structure near the water's edge, and often in a reed-bed. It is made of dead vegetation. The female usually lays four eggs which are incubated by both sexes for about a month. The young stay with the parents for 2 ½ months.

Little Grebe *Podiceps ruficollis*

The smallest European grebe, ranging throughout Europe and also found in North Africa, Asia Minor and parts of Russia. It lives on small as well as large lakes, preferring those with dense vegetation. The diet consists of crustaceans, aquatic insects, tadpoles and small fishes. This is a shy, retiring bird which is not often seen. The floating nest is built of dead plants, usually in dense vegetation and only a few metres from the bank. The female lays a clutch of 4-6 eggs between the beginning of May and July. These are incubated by both parents for about three weeks. The chicks remain with the parents for 6-7 weeks and when small they are sometimes seen on the back of one of the parents; this also happens in the Great Crested Grebe.

Bittern *Botaurus stellaris*

A widely distributed species in Europe, Asia, Africa and Australia, living mainly in the reed-beds bordering large lakes. In Britain this is a winter visitor, except that it breeds regularly in East Anglia, and sometimes in Kent and Lincolnshire. It has a characteristic dull booming call. When alarmed the long neck is extended upwards and the vertical stripes in the plumage blend in with the surrounding background of reeds, so that the bird becomes almost invisible. Bitterns feed on insects, crustaceans, fishes, frogs, newts, small birds and mammals. The nest is a pile of old reed stems and the female lays 4-5 brownish eggs. These are incubated by the female alone for 25-26 days. The young start to leave the nest 2-3 weeks after hatching and climb about among the reeds. They are fully fledged when about 8 weeks old.

The Little Bittern (*Ixobrychus minutus*) has a similar distribution and has possibly bred in East Anglia.

Heron *Ardea cinerea*

A common bird in many parts of Europe, Asia and North Africa on inland waters and along the coasts. The head has a long pendant crest. It is characteristic that when in flight Herons carry the head held back so that the neck is in the form of an S curve. Their main food consists of fishes which they spear with the bill, in shallow water. They also take insects, frogs and mice. Herons nest in colonies in trees. The clutch of 3-5 blue-green eggs is incubated by both parents for about 25 days. The young, which are also fed by both parents, become fully fledged when they are about 50-55 days old. Herons breed in many parts of Britain but are most abundant in the south-eastern counties.

Although similar in general appearance cranes and storks fly with the neck extended forwards.

Mute Swan *Cygnus olor*

Distributed throughout many parts of Europe and northern Asia, breeding in all parts of Britain, except the Shetland Islands. They can be seen on many large lakes and rivers. The distinctive black knob at the base of the otherwise orange bill is larger in the male or cob than it is in the female or pen. The knob is only seen in the adult birds. The diet consists of aquatic plants mostly taken from the bottom, and also insects, worms, frogs and small fishes. It has been suggested that Mute Swans may be responsible for the disappearance of areas of reed-bed, as they use reeds for the construction of their large nests. The 5-8 eggs are incubated, mainly by the female, for 35-38 days. At this time the cob is particularly aggressive in guarding the nest. The young, known as cygnets, are fledged at about 4½ months.

Grey Lag Goose *Anser anser*

Breeds in large areas of Europe and northern Asia. In Britain this species is seen mainly as a winter visitor, but small numbers do breed in certain parts of northern Scotland. It is distinguished from other goose species by the orange bill and the flesh-coloured legs and feet. It feeds on grasses and other plants. Grey Lags become sexually mature at the end of their 2nd year. They build a large nest of reeds and other plant material. The female lays 4-6 eggs, although clutches with up to 9 eggs have been recorded. The eggs are at first pure white but they become soiled during the course of incubation. They are laid at the end of March or the beginning of April and are incubated by the female alone for about four weeks. The young become fledged when they are about ten weeks old.

Grey Lag Geese are, of course, the ancestral stock of ordinary domestic geese.

Mallard *Anas platyrhynchos*

A very common breeding bird in Europe, Asia and North America, and the ancestor of ordinary domestic ducks. Mallard belong among the dabbling ducks, a group which also includes Teal, Garganey, Gadwall, Wigeon, Pintail and Shoveler. These ducks feed by dabbling in shallow water, up-ending so that the rear end of the body sticks out of the water. Dabbling ducks have an elongated body. The drake or male Mallard (above, left) has a strikingly coloured plumage whereas the female (above, right) is more inconspicuously coloured. This is associated with the fact that the female alone incubates the eggs and must be well camouflaged when sitting on the nest. Breeding territories may be established as early as the end of February. The nests are well hidden among dense vegetation close to the water's edge. The female lays 7-10 pale green eggs which are incubated for about four weeks. The young birds are fledged in about 7½ weeks.

Teal *Anas crecca*

A breeding bird throughout almost the whole of Europe, including Britain, and northern Asia. There is a similar form of Teal in North America. During the greater part of the year (October-July) the drake has a characteristic chestnut head with a green stripe over each eye and extending back to the nape. From July-October the drake has a mottled brownish plumage similar to that of the female, but the upperparts and breast are darker. Teal feed on worms, insects, snails, as well as water plants and seeds. They nest on the ground in May-June and the female lays 8-10 pale yellowish eggs which she incubates alone for 21-23 days. The nest is lined with down which is pulled over the eggs when the female leaves the nest. This serves to hide the eggs from flying predators. The young birds are tended by the female, who is assisted by the male, and they are fledged after about 24 days.

Pochard *Aythya ferina*

A breeding bird in Europe and northern Asia, and commonly seen in Britain where it breeds in the eastern part of the country. This is one of the diving ducks, a group which also includes the Tufted Duck, Goldeneye, Scaup, Red-crested Pochard and Ferruginous Duck. These obtain their food by actually diving to the bottom. They can therefore search for food in deeper water than the dabbling ducks which can only feed in shallow water near the banks. Diving ducks differ in body shape from the dabbling ducks for they have to be adapted for movement under the water; they are shorter and more flattened and the tail usually lies along the water surface. In addition, diving ducks swim underwater using their feet and not their wings as auks and penguins do (and also the marine ducks such as the Eider). From October to June the drake Pochard wears breeding plumage as illustrated. The female is dark brown with greyish back and flanks, the throat and the sides of the head being greyish-white. She lays a clutch of 7-11 greenish-grey eggs which she incubates alone for 24-28 days. The young are fledged when they are about 50 days old.

Tufted Duck *Aythya fuligula*

Distributed in Europe and northern Asia, this is a common duck in Britain where it breeds in most parts. In breeding plumage, during the winter, spring and summer the drake is black with conspicuous white on the sides, but the female is very dark brown. Tufted Ducks take some plant food but feed mainly on insects, molluscs, frogs, small fishes and spawn. The nests are well hidden among vegetation near to the water, often on a small island. The female lays a clutch of 6-14 eggs, usually in May, which she incubates alone for 23-25 days. She also tends the ducklings which are fledged at about 6 weeks.

Goldeneye *Bucephala clangula*

A breeding bird in much of Europe, northern Asia and North America, but in Britain only recorded as a migrant and a winter visitor. Drakes in breeding plumage have black upperparts, a shiny blackish-green head and white underparts. The white patch on the face is quite characteristic; the eyes are golden-yellow. Females, on the other hand, are mottled grey with a brown head, a white collar and a dark grey bill. Goldeneyes feed mainly on animal food (crustaceans, snails, insects) but take some plant food during the autumn. They nest in natural holes, sometimes those made by woodpeckers, or in rabbit holes; in Scandinavia they have nested in nest boxes suspended close to the edge of a lake. The clutch of 8-11 eggs is incubated by the female alone for about a month. Soon after hatching the ducklings are enticed by the female's call to leave the nest and jump down to the ground. They are fledged when about 8-9 weeks old.

Goosander *Mergus merganser*

Widely distributed as a breeding bird in Europe, northern Asia and North America, in Britain nesting at several localities in Scotland and a few places in northern England. Goosanders feed primarily on fishes, as well as frogs, insects and crustaceans. They are sexually mature in the second year. They nest in natural holes or among boulders in the vicinity of water. The female lays a clutch of 7-12 eggs which she incubates alone for up to five weeks. The young birds remain in the nest for 2-3 weeks and then continue to be tended for a time by the female. During this period they may be transported on the female's back. They are fledged at an age of 60-70 days.

This species and the Red-breasted Merganser (*Mergus serrator*) and the Smew (*Mergus albellus*) belong to a group known as the saw-billed ducks.

Marsh Harrier *Circus aeruginosus*

A breeding bird in Europe and western Asia. In Britain mainly known as a summer visitor, although a few pairs breed in East Anglia and there are records of breeding in North Wales and southern England. Harriers are hawks with long legs and long wings and they fly low over the ground with slow wing beats alternating with periods of gliding. The male (illustrated) has brown plumage with pale grey on the tail and part of the wings, whereas females are a more uniform dark brown with a yellowish head and shoulders. The nest is built by the female among reeds and other coarse vegetation. The eggs, usually 4-5, are laid in late May or early June and are incubated mostly by the female for 30-38 days. The chicks remain in the nest for about five weeks and are fully fledged after a further three weeks. Marsh Harriers feed on frogs, birds, eggs and small mammals.

Osprey *Pandion haliaëtus*

One of the most widely distributed birds, with races in Europe, Asia, North Africa, North America and Australia. Seen regularly in Britain as a passage migrant, and after an interval of almost fifty years now breeding on a very limited scale in Scotland. The populations in Europe were drastically reduced owing largely to extensive egg collecting, but they are now protected. Ospreys feed almost entirely on fishes. The nest is built on cliffs, ruins or sometimes in a pine tree. It consists largely of sticks, moss and seaweed. In late April or early May the female lays 2-3 eggs which are incubated, mainly by her, for about five weeks. The chicks are fed, usually by the female, on food brought by the male. They are able to fly when about 8-10 weeks old.

Coot *Fulica atra*

A common breeding bird in Europe, temperate Asia and North Africa, and there is also a race in Australia. It breeds throughout Britain, except in the Shetlands, and is commonly seen on lakes and ponds, even in towns. During the winter Coots often congregate in large flocks. The diet consists mostly of plant food, including the soft shoots of aquatic plants which they get by diving. They also take some worms, snails, insects and fishes. Both sexes build the large nest, using the dead leaves of reeds and other waterside plants. In late March or early April the female lays 6-9 eggs, sometimes as many as 15. These are incubated by both the parent birds for 21-24 days. The young leave the nest after 3-4 days, but return to it at night. They start to dive and search for their own food when they are about four weeks old and are usually independent at eight weeks.

Moorhen *Gallinula chloropus*

A common breeding bird in Europe and Asia, with races in other parts of the world. Breeds in all parts of Britain, usually close to lakes and ponds. The nest is built by both sexes in dense vegetation close to the water's edge. It consists mainly of dead sedges and reeds. The eggs, usually 5-11, are laid in March-April, occasionally at other times, and are incubated by both sexes for about three weeks. The young are fledged at about 7-8 weeks. It is quite common for a pair to raise two broods in the year, and sometimes three.

Coots and Moorhens belong to the rail family, which also includes the Water Rail (*Rallus aquaticus*) and the Spotted Crake (*Porzana porzana*).

Little Ringed Plover *Charadrius dubius*

A breeding bird throughout Europe and many parts of Asia. A summer visitor in Britain, which in recent times has bred in small numbers in England north to Yorkshire and also in Cheshire and Gloucester. European populations spend the winter in the Mediterranean area and further south. They feed mainly on insects, worms, spiders and molluscs. The nest is merely a shallow depression in sand or shingle, sometimes lined with fragments of shell or dead plants. The female lays 3-5 (usually 4) eggs in May and these are incubated by both sexes for just over three weeks. The young are fledged at 24-30 days. The very similar Ringed Plover (*Charadrius hiaticula*) has a conspicuous white wing bar seen when the bird is in flight, but this is absent in the Little Ringed Plover. The pattern of markings on the head is also slightly different.

Common Sandpiper *Actitis hypoleucos*

Distributed throughout northern and central Europe, except Iceland, and northern Asia. In Britain this species breeds in Scotland and northern England southwards to Hereford, and at a few places in Devon and Somerset. The diet consists mainly of worms, crustaceans and insects with some plant matter. The nest is a hollow lined with a few grass stems, and is made close to water, often on a shingle bed, but usually in among plants. They have been known to nest low down in a tree. Between the middle of May and early June the female lays 3-5 (usually 4) eggs which are incubated by both sexes for about three weeks. The young birds are tended by both parents and they are fledged at about four weeks.

Sandpipers and plovers belong among the large group of birds known as waders or Charadrii.

Black-headed Gull *Larus ridibundus*

A very common breeding bird in Europe and parts of Asia. In Britain there are breeding colonies in most of the coastal counties and also in some inland areas. Some of these breeding birds migrate to the south in early autumn, but others are sedentary. These small white gulls are particularly abundant in certain towns and cities, including London. In the breeding plumage (illustrated) the head is chocolate-brown and the leading edge of each wing is white. In winter the dark head coloration is lost but there is still a dark patch in the ear region. The nests are usually made on the ground among vegetation. Black-headed Gulls feed on a wide variety of animal and plant food, but particularly crustaceans, molluscs, insects, earthworms, small fishes and even mice. Plant food includes grasses, seeds, seaweed, potatoes; they also scavenge for scraps of bread and meat. From the middle of April onwards the female lays a clutch of 2-6 (usually 3) eggs. These are incubated by both sexes for about 23 days. After hatching the chicks are fed by both parents and are fledged when they are 5-6 weeks old.

Common Tern *Sterna hirundo*

Widely distributed in various similar races in Europe, Asia, North Africa and North America. Breeds on many coasts of Britain, and often appears inland when on migration in spring and autumn. Terns are more delicately built than gulls, and their flight is much more graceful. They feed on worms, crustaceans, molluscs, insects and small fishes. Some of the food is picked up from the surface when the bird is in flight. The nests are in colonies. Each nest is merely a shallow depression on sand or a shingle bank or sometimes on a small islet. In late May or early June the female lays a clutch of 3 eggs, sometimes 2 or 4, and these are incubated by both sexes for about four weeks. The chicks are fed by both parents until they are fully fledged about three weeks after hatching. They may even be fed by the parents for a further week.

Reed Warbler *Acrocephalus scirpaceus*

A breeding bird in Europe and north-western Africa, and seen in Britain as a summer visitor which breeds in many English counties but is only known as a rare vagrant in Scotland. The warblers of the genus *Acrocephalus* are small birds mainly associated with standing waters and the Reed Warbler is one of the commonest. The Sedge Warbler (*Acrocephalus schoenobaenus*) differs in having a conspicuous cream-coloured eye stripe and dark streaks on the crown and mantle. The Great Reed Warbler (*Acrocephalus arundinaceus*) breeds on the continent of Europe but is only seen as a rare vagrant in Britain, mainly in south-eastern England.

The Reed Warbler lives in marshy areas, particularly in reed-beds or in bushes close to water, and more rarely in parkland and gardens. It spends most of its time in among vegetation and is particularly adept at climbing among reed stems and flitting ceaselessly from one to another. In this habitat it never flies very far. The diet consists mainly of insects, but it also takes spiders, small snails and slugs, with some berries during the autumn. This and other warblers have a characteristic churring song, and only the expert can distinguish between the songs of the different species. The nest is a well-built structure of grasses and reeds with a deep cup which is lined with hair, wool, feathers and reed flowers. It has to be flexible enough to withstand the wind. The female lays a clutch of 3-5 but usually 4 eggs. Both sexes take turns to incubate the eggs which hatch after about 11 days. The chicks are fed by both parents and are fledged at about 12 days. The illustration shows a Reed Warbler rearing a young Cuckoo, which happens quite frequently.

Dipper *Cinclus cinclus* (above, left)

Distributed throughout most of Europe and temperate Asia, and breeding in most parts of Britain, except south-eastern England. Dippers live on fast-flowing rivers and streams and along the edges of lakes in mountainous areas. They hunt for food in the water, for they dive and swim well. They actually walk along the river bed searching for worms, crustaceans, molluscs, tadpoles and small fishes. The nest is built on a cliff ledge, among trees or on the brickwork of a bridge. It is a cup-shaped structure of dry grass and moss, lined with dead leaves. The 3-7, usually 5, eggs are laid in March or April and are incubated by the female alone for about 16 days. The chicks are then fed by both parents and are fledged at about 24 days.

Kingfisher *Alcedo atthis* (above, right)

A breeding bird throughout most of Europe and temperate Asia, except Iceland and most of Scandinavia. Breeds in most parts of Britain, except north-west Scotland. Kingfishers live close to rivers, canals and lakes, and in winter sometimes on rocky shores. They perch on a branch overhanging the water, keeping watch for fishes and aquatic insects, which they catch by rapidly diving. The nest is a bare chamber at the end of a tunnel bored in a steep river bank. Both parents incubate the 6-7 eggs which hatch in about 20 days. The young, which are fed by both parents, are fledged at about 24 days.

Grey Wagtail *Motacilla cinerea* (below)

A common breeding bird throughout Europe, except Scandinavia, and extending eastwards into temperate Asia. Breeds in most areas of Britain. The popular name refers to the constant vertical flicking movements of the tail. Wagtails feed mainly on insects, but also take some small snails. They nest on the ground, in a depression among dense vegetation. The female lays 5-6 eggs in May or early June and these are incubated by both sexes for about 12 days. The chicks are fed by both parents and are usually fledged after a further 12-13 days.

Water Shrew *Neomys fodiens*

Distributed throughout Europe, except most of the Iberian Peninsula, Ireland and many areas in the Balkans, but extending eastwards across the whole of temperate Asia. It breeds throughout Britain. This and other shrews, as well as the Hedgehog and Mole, belong among the insectivores, and are not related to the rodents (rats, mice, voles etc.). Water Shrews, which are c. 10 cm long, dive and swim well. They are mainly active at night, searching for worms, snails, aquatic insects and their larvae, frogs and small fishes. They live in underground tunnels. After a gestation period of about 24 days, the female gives birth, in April-May, to a litter of 3-8 young, which each weigh about 1 g. These are suckled for about five weeks, and are sexually mature in the following summer. The female may produce 2-3 litters during a single summer.

Musk-rat *Ondatra zibethicus*

A rodent native to North America, which was first introduced into Europe in 1905, and is now feral in many areas. It came to Britain about 1929, and colonies were established in many places; it became a serious pest and was finally exterminated in 1937. Although rather slow-moving on land, Musk-rats swim very well, using the hind-limbs and tail for propulsion. They feed mainly on the shoots and leaves of water plants, in winter on roots and some animal food, including fishes. They build burrows in steep banks, up to 10 m long. The nest, lined with grass and moss, also forms part of the underground tunnel. In low, swampy areas they may build "lodges" of twigs and grasses, not unlike those of Beavers, each of which is occupied by a single pair and their young. The female gives birth to her first litter in April-May, usually with 6-8 young. She may produce 2-3 more litters during the summer and autumn.

Coypu *Myocastor coypus*

A rodent native to South America, and known in the fur trade as nutria. It has been introduced into Europe, including the U.S.S.R., and escaped individuals from fur farms have now established feral populations in several countries including England. Coypus are about the size of a hare, and they weigh 6-9 kg. They live along the banks of lakes and slow-flowing rivers, where they are mainly active at night. They move rather clumsily on land but dive and swim very well. They feed primarily on water and marsh plants, also freshwater mussels, and in some areas damage agricultural crops. They dig subterranean tunnels in the banks. They can breed at any time of the year, and there may be two litters annually. Each litter has about five young, which are born after a long gestation period of about 130 days. The young are born with their eyes open and fully furred, and they can swim within a few hours.

European Beaver *Castor fiber*

A large rodent up to c. 130 cm long, including the broad, flat, scaly tail. At one time very widespread in Europe, the Beaver is now found only in scattered populations in Norway, Sweden and Finland, in the lower reaches of the Rhone and in one area on the Elbe. The Beaver is well known for its ceaseless activity in damming rivers to produce calm lakes in which it builds a lodge to live in. The dams and lodges are built of branches held together with mud. The lodges are sometimes up to 4 m in diameter and up to 2 m in height. The animals use their large powerful teeth to cut down trees to provide the necessary timber. They feed mainly on the bark, leaves and shoots of Birch, Willow and Aspen and on various other plants. The young are born after a gestation period of about three months and there are usually 3-4 in a litter. These remain with the parents until they become sexually mature at an age of three years.

Otter *Lutra lutra*

A largely aquatic mammal widely distributed throughout Europe, except Iceland and many of the islands in the Mediterranean. Otters are native to Britain and Ireland, although their numbers have much decreased in recent years, particularly in parts of England. They prefer clean standing and flowing waters, especially streams and rivers whose banks offer shelter and hiding-places. They occur in mountains up to altitudes of 2500 m. It is not difficult to see how water pollution and consolidation of the banks has restricted their environment.

Otters are up to 80 cm long and they weigh up to 15 kg. It is characteristic that their front and hind feet are webbed. However, they swim mainly with the tail, helped by the hind feet. They are primarily nocturnal, although they may be active during the day in undisturbed places. They carry out regular sorties along their own stretch of water and also on land. It has been recorded that they move upstream during the morning or sometimes during the day, returning downstream at night. This behaviour pattern is associated with the fact that they excavate various dens along their stretch of water, which they visit from time to time. Such dens, or holts as they are called, usually have an entrance underwater and a ventilation shaft. In addition, they also have hiding-places in hollow trees and natural holes. They feed mainly on fishes, but also take frogs, aquatic birds and rodents, as well as crustaceans.

The biology of the Otter is still relatively poorly known, although in recent years considerable progress has been made in this field. They are now being kept and bred under almost natural conditions in several wildlife parks, and there is a possibility that some of these can be re-introduced into the wild. Otters apparently have no fixed mating season for young individuals can be seen throughout the year. After a gestation period of about 60 days the female gives birth to 2-4 young, which are independent at 6-9 months, although they remain with the mother for a further period.

Index

English Names

Latin Names